ANCIENT MAN IN AMERICA.

INCLUDING

WORKS IN WESTERN NEW YORK,

AND PORTIONS OF OTHER STATES, TOGETHER WITH STRUCTURES

IN CENTRAL AMERICA.

———————

BY

FREDERICK LARKIN, M. D.,

Member of the American Association for the Advancement of Science.

———————

PUBLISHED BY THE AUTHOR.

1880.

MILES A. DAVIS, printer.

ISBN 1-60135-486-X

J. Larkin

PREFACE.

In presenting this book to the public I entertain a hope that it may prove worthy of perusal, especially to that class who have not the means to reach the large and more elaborate works which have been written on the subject of American Antiquities.

I had no intention of publishing this volume until within the last few months. About the first of January, 1879, I commenced a series of articles which were published weekly, in a newspaper— the RANDOLPH REGISTER—describing the mounds and fortifications which have been explored by myself and others, on the banks of the upper Allegany River and some of its tributaries.

The Conewango Valley, one of the most extensive in the State of New York, has disclosed a great number of ancient works, such as sepulchral mounds and what is supposed to have been military fortifications, composed of a ditch and parapet designed for protection against a hostile enemy. As these works have never been elaborately described by any other writer and as they formerly contained many curious and interesting relics peculiar to that ancient and wonderful people, (the mound builders), my papers attracted much attention and were sought for by several archælogical societies and published in whole or in part in some forty prominent newspapers.

The favor which my small contribution to that interesting and important subject, archæology, has received from the press and from other directions, has induced me to re-write and present in a more elaborate form some of the most prominent and interesting ancient works which have been found in Western New York, and also in the Ohio and Mississippi Valleys, together with those found in the vale of Mexico and in parts of Central America.

The theory that the ancient works located in Western New York were those of the modern Indians I consider erroneous and have endeavored to present evidence that those mounds and fortifications were the works of the Mound Builders, or more especially those which contain copper relics and mica.

What I have written in relation to Central America I regard as authentic, for I have endeavored in all cases to consult the most reliable authorites.

In accordance with my design in preparing this work for the press, I have endeavored to present it in a style that will interest the popular mind. I am well aware that it will add but a trifle to the great number of elaborate works already before the public; although what is presented in regard to many of the ancient works and relics found in Western New York which have been explored only by myself, must necessarily interest those in search of new discoveries.

My theory that the pre-historic races used, to some extent, the great American elephant, or mastodon, I believe is new and no doubt will be considered visionary by many readers and more especially by prominent archæologists. Finding the form of an elephant engraved upon a copper relic some six inches long and four wide, in a mound on the Red House Creek, in the year 1854 and represented in harness with a sort of breast-collar with tugs reaching past the hips, first led me to adopt that theory. That the great beast was contemporary with the mound builders is conceded by all, and also that his bones and those of his master are crumbling together in the ground.

Nearly all of the great works of the ancient inhabitants of America which have withstood the ravages of time, have a religious significance, such as sepulchral and truncated mounds, and also many of the great works found in Central America; hence in my remarks in regard to those wonderful works I have treated many of them as having been the product of the religious element which is one of the strongest in the nature of man.

Many of the mounds and fortifications which I have attempted to describe are now nearly obliterated but have been presented as they appeared when first surveyed by the white man. The Indian races who occupied this country subsequent to the mound builders, I believe never disturbed the tumuli to any great extent, though they frequently buried above the remains of the ancient dead.

Notwithstanding that I have written this work to interest the popular reader and without any correct order of arrangement, I have endeavored in all cases to present facts in regard to the peculiar characteristics of the works of which I have treated, and have taken great pains to consult the most reliable authorities and epecially those who have made personal examinations. I am well aware that many popular writers in describing the ancient works have been frequently mislead by conflicting reports which are always in circulation when connected with any subject that borders on the marvelous.

I have myself examined many ancient works found in the Ohio Valley and other localities which have been incorrectly described, and many visionary theories presented with regard to the purpose for which they were constructed.

If prominent archæologists have adopted wild and irrational theories with regard to the ancient inhabitants and their works, it is not improbable that an individual like myself should fall into errors while endeavoring to interpret some of the mysteries connected with the works left by the inhabitants of ancient America.

While collecting material for that part of this work confined to Western New York, I am indebted to the Hon. Obed Edson, of Sinclairville,

Judge L. Bugbee, of Stockton, and Charles L. Bishop, of Jamestown, for valuable information and assistance.

For a description of the ancient works located in Ohio and the Western States I have consulted various writers, but have drawn much information from papers written by Rev. J. P. McLean, a gentleman who has explored much of that country, and made a personal examination of the works of which he has given a description.

For information with regard to Mexico and Central America, I have consulted Stephen's Travels, Baldwin's Ancient America, and other works.

The illustrations prepared for this work representing mounds and fortifications found in Western New York I know to be authentic, as I have given them a personal examination. Most of the engravings representing works in Ohio, Wisconsin and Central America were made from illustrations formerly published in the Smithsonian reports, for the right of which I am indebted to the courtesy of Prof. Baird. All the engravings were made expressly for this book, but as the work was done by a person not thoroughly versed in the art I am aware they can be justly criticised.

F. LARKIN.

Randolph, N. Y., 1880.

CONTENTS.

ANCIENT MAN IN AMERICA.

CHAPTER I.

DISCOVERIES IN WESTERN NEW YORK.

A simple heap of stones or earth seems to have been the first monument that suggested itself to man. In the old world thousands of such monuments are known to exist. The great antiquity of those works has staggered the greatest minds of Europe. By whom, and when built, remains an impenetrable mystery. It is the opinion of modern archæologists, that the earthworks found scattered over most parts of the old world, were built thousands of years before the historic period, and that to the builders of the splendid cities of Baalbec and Palmyra, they were as much of an enigma as to the inhabitants of the nineteenth century. But rude as these primitive memorials are, they have been little impaired by time, while majestic and imposing structures have fallen into shapeless ruins. "When covered with forests, and their surfaces interlaced with the roots of trees, or when protected by turf, the humble mound bids defiance to the elements which throw down the temple, and crumble the marble into dust."

No part of the world presents to the archæologist so many remarkable works as the Western Continent. We can trace the ancient inhabitants from the Northern Lakes through the Allegany, Ohio and Mississippi valleys, to Mexico, Brazil, Peru, and the Pacific Ocean. Every hill, mountain and valley discloses some rude instrument used by a people whose history is enveloped in midnight darkness. So numerous was this people in the Mississippi Valley, that Mr. Brackenridge, after extensive explorations, says: "There are traces of a population far beyond what this extensive and fertile portion of the continent is supposed to have possessed; greater, perhaps, than could be supported by the present white inhabitants, even with the careful agriculture practiced in the most populous parts of Europe."

It was formerly supposed that there was some alliance with those people and the Indian races of America, but that is very improbable. The skull of the Indian bears but little resemblance to that of the mound-builder. The head of the Indian is low in the moral sentiments, has large firmness and small benevolence; in fact his head indicates the cruel savage that he is. The mound-builder has a head that will compare favorably with the most intellectual people now living. His head is well developed in the moral and intellectual regions, with large firmness which caused him to move with energy, whether in peace or in war.

When the copper mines of Lake Superior began to be explored, traces of ancient works were found

in almost every section. In several places shafts
had been sunk to the depth of thirty feet or more,
and veins of native copper were traced for a great
distance, in many places through solid rock. In
one of the shafts was found a number of tools made
of copper, which, at the time of the discovery, were
reported as being hardened at one end; if such
was the case, it was done with tin or with some
other process unknown to us. From the shores of
Lake Superior we can trace this people to Wisconsin, where we find some singular earthworks : six
effigies of animals, six parallelograms, one circle,
and one effigy of the human figure. These tumuli
extend for the distance of half a mile along the
trail. ·What the animals represent in effigy is difficult to determine. Many at the present time suppose that the mastodon is one, and that he was a
favorite animal and perhaps used as a beast of burden. That the mastodon was contemporary with
the mound-builders is now an undisputed fact. It
is a wonder, and has been since the great mounds
have been discovered, how such immense works
could have been built by human hands. To me it
is not difficult to believe that those people tamed
that monster of the forest and made him a willing
slave to their superior intellectual power. If such
was the case, we can imagine that tremendous teams
have been driven to and fro in the vicinity of their
great works, tearing up trees by the roots, or marching with their armies into the field of battle amidst
showers of poisoned arrows. In Western New
York, particularly within the borders of the great

valley of the Conewango river, evidence is abundant that ancient man and the great American elephant trod the soil together. Large molar teeth have been found at East Randolph, Leon, Connewango, Ellington and various places near the tributaries of the Conewango.

From northern Wisconsin we can trace these ancient people through Michigan and Ohio to the southern shore of Lake Erie. Near the mouth of Cattaraugus Creek, in Western New York, commences a series of works which extend through the Conewango valley. In the town of Dayton are the remains of a sepulchral mound in a circular form, one hundred and twenty feet in circumference, with an elevation of ten feet. This tumulus, when explored, was found to contain several skeletons which were no doubt those of some distinguished warriors whose acts of heroism might have been connected with the decline and fall of that powerful race of men. On either side of the valley a chain of fortifications can be traced for more than twenty miles, interspersed with numerous mausoleums constructed to cover the remains of their distinguished dead.

In the towns of Leon and Conewango a number of burial mounds have been explored. Among them is one that contained eight skeletons which were buried in a sitting posture, in such a manner as to form a circle within the mound. In the center of this circle, surrounded by the mouldering skeletons, stood a pestle artistically wrought from granite. This relic was placed in a perpendicular

position and encircled with twenty-four flint arrows of large dimensions. This mound is now
nearly obliterated, and the ground whereon it once
stood is cultivated by the white man. The fields
in either direction disclose large quantities of relics designed for warlike purposes, which have been
discharged, no doubt, during some terrible battle.

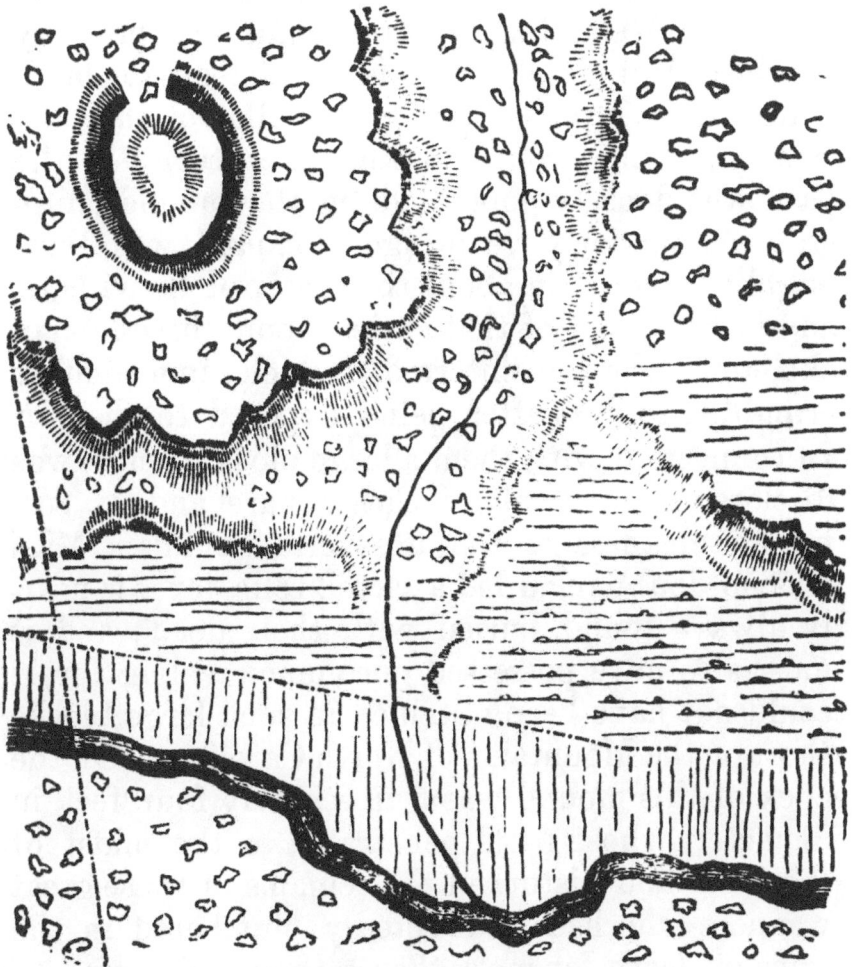

Fig. 1.—Works in Conewango.

TUMULI AND ANCIENT WORKS SOUTH OF THE
CONEWANGO.

On the Little Conewango, about two miles from
its junction with the main stream, is located the
village of Randolph. The ground occupied by
this beautiful village is a level plateau of land,
which forms a part of the great valley of the Conewango. This ground, unknown centuries in the
past, was densely populated by the mound-builders. About forty years ago a mound was thoroughly explored in about the center of the village,
and in the midst of that ancient tomb three human
skeletons were disinterred, which immediately
crumbled to dust after being exposed to the air.
In connection with them a large block of mica was
disclosed about twelve inches square and of sufficient thickness to weigh several pounds. Mica
was no doubt regarded a sacred relic, for it is nearly always found buried with their dead; and it
must also have possessed uncommon value, for it
could not have been obtained in such large blocks
short of the Mountains of North Carolina. On the
apex of this mound was a tree nearly four feet in
diameter, whose roots penetrated to the midst of
the tomb and disturbed the remains of those great
warriors who had undoubtedly slumbered in the
arms of death for more than a thousand years.

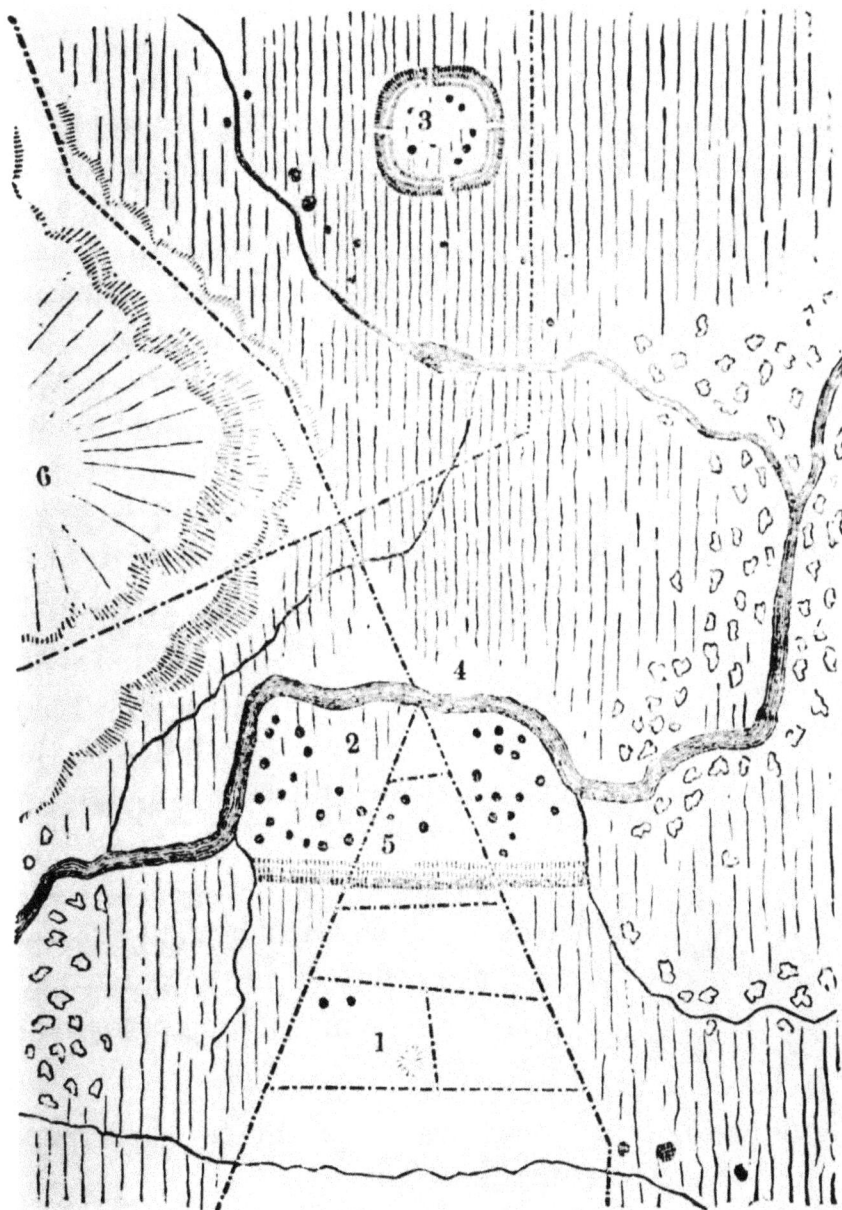

Fig. 2.—Plot of Randolph Village. Fig. 1 shows the tumulus located in the village; 2, numerous caches near the little Conewango; 3, Military fortification; 4, the Little Conewango; 5, redout sixty rods in length; 6, location of the Chamberlain Institute.

It was the opinion of Dr. Cheney, with whom I have spent much time in exploring the mounds and ancient works in the vicinity of Randolph — and my own coincides with his — that the ground now occupied by the village was once an ancient city; for when the white man came he was astonished to find within its borders and adjacent to it the remains of military fortifications, mounds, caches, hearths, and various relics used for domestic and warlike purposes. In the summer of 1878, a Mr. Scudder while plowing a piece of ground, about forty rods from the Railroad Depot, discovered a nest of arrow heads under the remains of a large pine stump, and, by exploring, unearthed 167. Subsequent to that time about fifty more have been found together with blocks of mica, stone axes, &c. That these relics were buried before the growth of the tree is very apparent. The tree was one of the largest in the country. From as correct a measurement as can be made, it was about eight feet in diameter and was down from age long before the

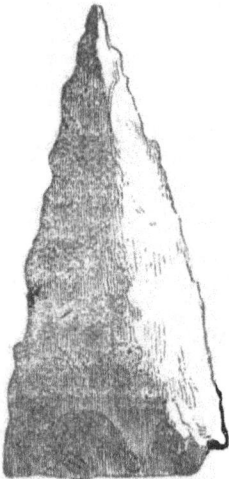

Fig. 3.—Form of this arrow.

axe of the white man was brought to bear upon the forest trees in Randolph.

On Elm Creek, near the village of East Randolph, is an embankment and trench of circular form two hundred and eighty feet in diameter.

From the appearance of a successive growth of
timber it would indicate a remote date of construction.   Within and adjacent to this work numerous
caches and hearths have been discovered, some of
which contained the remains of charred corn.   In
the vicinity, where the ground has been cultivated,
hundreds of relics used by those ancient warriors
have been found.

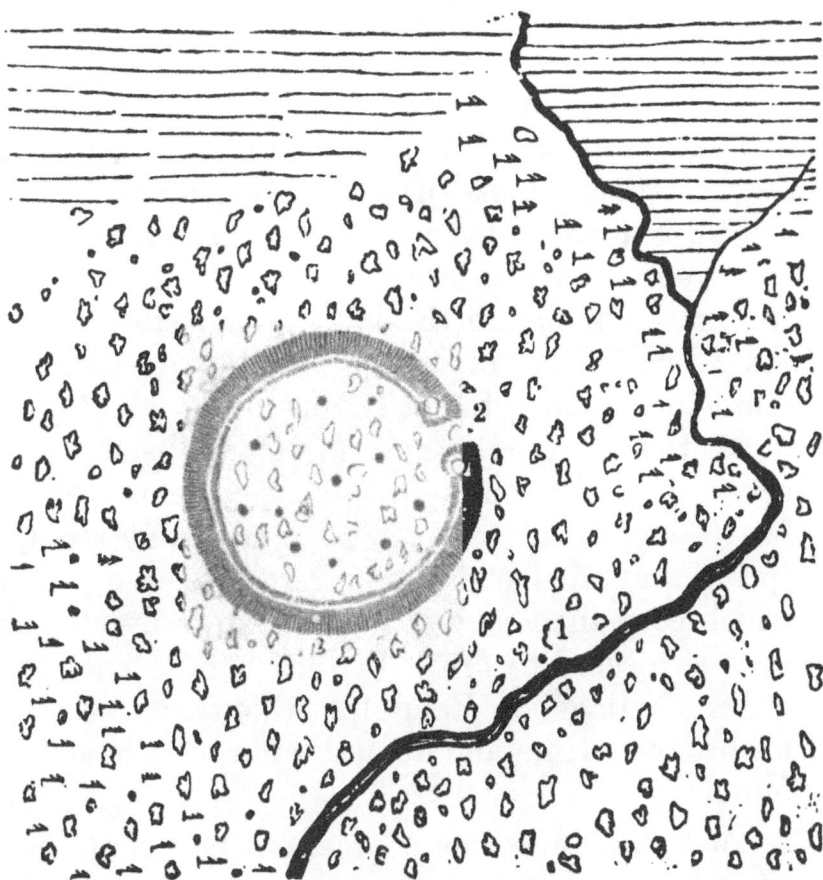

Fig. 1.—Ancient works on Elm Creek. Fig. 1 represents Elm Creek; 2, gateway through which the circle was entered; the round spots within the circle
represent caches.

Near the residence of J. E. Weeden, Esq., and about one mile from the village of Randolph, is a fortification that surrounds about three acres. The ditch was about eight feet deep, and numerous trees of great size were found growing upon the parapet and in the midst of the intrenchment. This work was well defined when first discovered in the wilderness, where it had lain in solitude for more than a thousand years, undisturbed only by the hurricane as it toyed with the monarchs of the forest.

About thirty years ago Dr. T. Apoleon Cheney, in a survey made of the village of Randolph, discovered the remains of a redoubt, sixty rods in length, that extended across the village plateau, on either end of which is a small rivulet fed by living springs. On the north side of Jamestown street Dr. T. A. C. Everet, while ditching his land and removing stumps, found blocks of mica and great quantities of relics, which had long been hidden in the ground from one to two feet below the surface. About six rods from a pure and singular spring a number of hearths were disclosed, one of which contained several loads of stones which had been subjected to fire. Within a mile surrounding the village of Randolph thousands of relics made for warlike and domestic purposes have been found.

That this numerous people, on their route from the northern lakes to the southern rivers, should build a city is very probable; and that the land now occupied by Randolph village was the place selected, is apparent from the vestiges found there. But if such a city once existed it ages ago crumbled to dust by the "effacing fingers of time." But the mighty arm of civilization has leveled the forest and on a beautiful eminence, on whose surface once sported these ancient people, can now be seen one of the most flourishing and ably conducted institutions of learning in Western New York. Here the god to whom those warriors lifted their prayers has become dethroned, and demonstrated to be the center of revolving worlds.

In reflecting upon these works and those who made them, the question comes to the mind who were the builders, and where are their descendants? But no satisfactory answer has ever been given.

An other race possessing different characteristics has subsequently appeared and for unknown centuries held undisputed empire over the American Continent. But they are fading away, save in the waning shadow of the western mountains, and even there "he withers before the strong arm of civilization, and soon his last footfall will die away."

Through all of the unknown centuries that have marked the ancient occupation of America, there has been found no annal, but only "the monuments now crumbling to ruin and dim in the twilight of ages."

We can trace this wonderful people from the Conewango Valley to that of the Allegany, and thence to the Ohio and Mississippi, where hundreds of their imposing structures still remain. But what part they played in the great drama of life is shrouded in the darkness of the tomb. Their only history is found in great piles of earth, military fortifications and crumbling skeletons.

Thousands of years ago our valleys and hills teemed with a great multitude of human beings cultivating the soil, hunting the deer and performing great deeds of valor. If Randolph and the adjacent country had a tongue, what a strange story could be told; great armies marching to and fro, mounted on the mastodon, that king of beasts, shaking the earth with his tread; soldiers driving their flint arrows into the quivering flesh of their victims; women with their children fleeing from the blazing rafters of their homes; priests praying to their fabled gods to make their wives widows and the children of the enemy fatherless; the tributary streams of the Conewango pouring down their turbid waters red with human gore; promises unfulfilled, vows unredeemed, and every element of virtue and benevolence trampled in the dust.

CHAPTER II.

TUMULI ON THE INDIAN RESERVATION.

All of the towns in Cattaraugus and Erie Counties are rich with relics left by the ancient people. On the north side of Cattaraugus Creek, on lands of the Indian reservation, is standing at the present time a group of mounds, which have never been disturbed save by the action of the elements and trees of the forest, which always attain an uncommon size when found among these tombs.

The most extensive of these tumuli is on the terrace, which is elevated about fifteen feet above the stream. It is oblong in shape, fifty-four feet in length by thirty-six in breadth and fifteen feet in height. Another tumulus, located about seventy rods in a southerly direction, is circular in form and about one hundred and twenty feet in circumference and sixteen feet in height. Other works of less interest to the archæologist are found in every direction.

The Indians of this reservation have a superstition with regard to disturbing these receptacles of the dead; they say their fathers desired them to remain undisturbed. Hence no one will be guilty of violating the sacred spot where the remains of some venerated chieftain is mouldering to dust. Not only in civilized, but in savage life, the desire to perpetuate the memory of the departed seems a powerful element indigenous to the nature of man.

Although those mausoleums have stood since a score of great forests have lived and perished, let them rest, for they will feast the eye of the antiquarian thousands of years after our generation shall have passed away.

> " How sleep the brave who sink to rest,
> With all their country's wishes blest—
> While Spring, with dewy fingers cold,
> Returns to deck the hallow'd mould. "

FORTIFICATIONS IN THE TOWN OF ELLINGTON.

In the town of Ellington, the relics and works constructed by the ancient inhabitants are very numerous. On a hill, about a fourth of a mile from the village, are the remains of a fortification which was no doubt constructed for defensive purposes. It occupies an eminence of more than a hundred feet above the waters of Clear Creek, and commands a view of the circumjacent country for many miles.

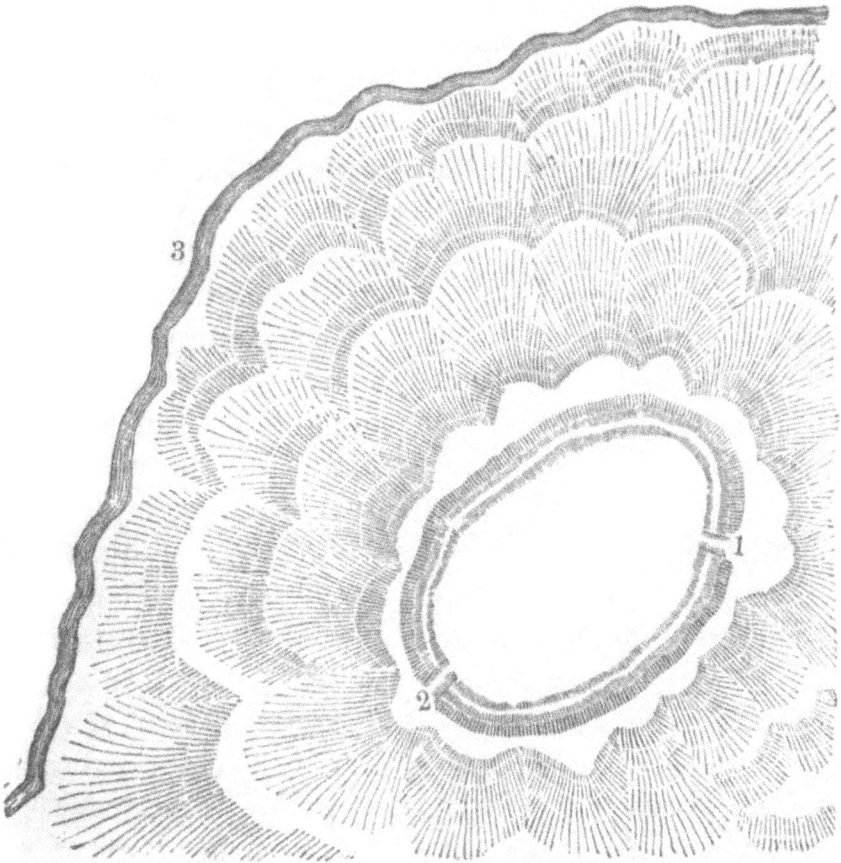

Fig. 5.—Great fortification in the town of Ellington, located on a hill. Figure 1 gateway leading towards the East; 2 entrance towards the South; 3 Clear Creek.

To the east, for several miles, is spread out the great valley of the Conewango, clothed with its native forest; to the westward is a broad ravine cut by the waters of a clear stream that rises among the hills which divide the valley of the Conewango from that of the Cassadaga. According to a survey made by Dr. Cheney, in 1868, this work is in the form of a parallelogram and is six hundred feet in length by two hundred and ninety in width.

The intrenchment, as it lay in the forest, was several feet deep. On the southern side was a wide gateway with an elevation on either side to guard the entrance. Within the intrenchment human bones have been disclosed in large quantities, and in such a stage of decay that they readily mingled with the soil. The land occupied by this fortification is noted for its great fertility, and marks the ground where once the tide of battle raged. This work presents more than ordinary interest to the antiquarian, on account of the great amount of implements of ancient warfare found in the vicinity.

About a mile and a half up the valley of Clear Creek, and near its junction with a mountain stream are two other works, about sixty rods apart.

One of these works presents a similar appearance to the one heretofore described. It stands upon an eminence, about seventy-five feet above the waters of a stream that flows from copious springs located amongst the hills. The vallum surrounds nearly two acres. It is in the form of a parallelogram; and, when first examined, the parapet was about three feet in height. The location of this work was well selected for defence against the near approach of an enemy, especially when armed with the implements used by those ancient warriors.

Near the summit of a hill, about half a mile west of the narrows, in the town of Napoli, is a tumulus about one hundred and twenty feet in circumference. From its elevated position — more than six hundred feet above the valley of the Conewango — it commands an extensive view of the surrounding country for several miles in either direction. This mound has never been thoroughly explored; but a few years ago a small amount of soil was removed near its base, which disclosed some relics, among which was a flint spear head more than six inches in length and wrought with remarkable skill. This mound, no doubt, contains the remains of some venerated dead who have slept for unknown centuries in the wild solitude.

About eighty rods in a westerly direction from this tumulus is a rapid stream that winds its way to the Allegany River. As we cross the river and pass up the Red House valley, on an eminence, about two hundred feet above the level of the stream, we find an ancient fortification from which can be seen the surrounding hills and the waters of the great river. This work is very extensive, being one thousand and sixty feet in circumference. The parapet is about three feet in height, and the ditch two feet in depth. Two rapid streams which rise in the deep ravines nearly encircle the eminence upon which this work is located. Near the centre of the inclosure is a copious spring, which was selected to supply the ancient warriors with water during a protracted siege.

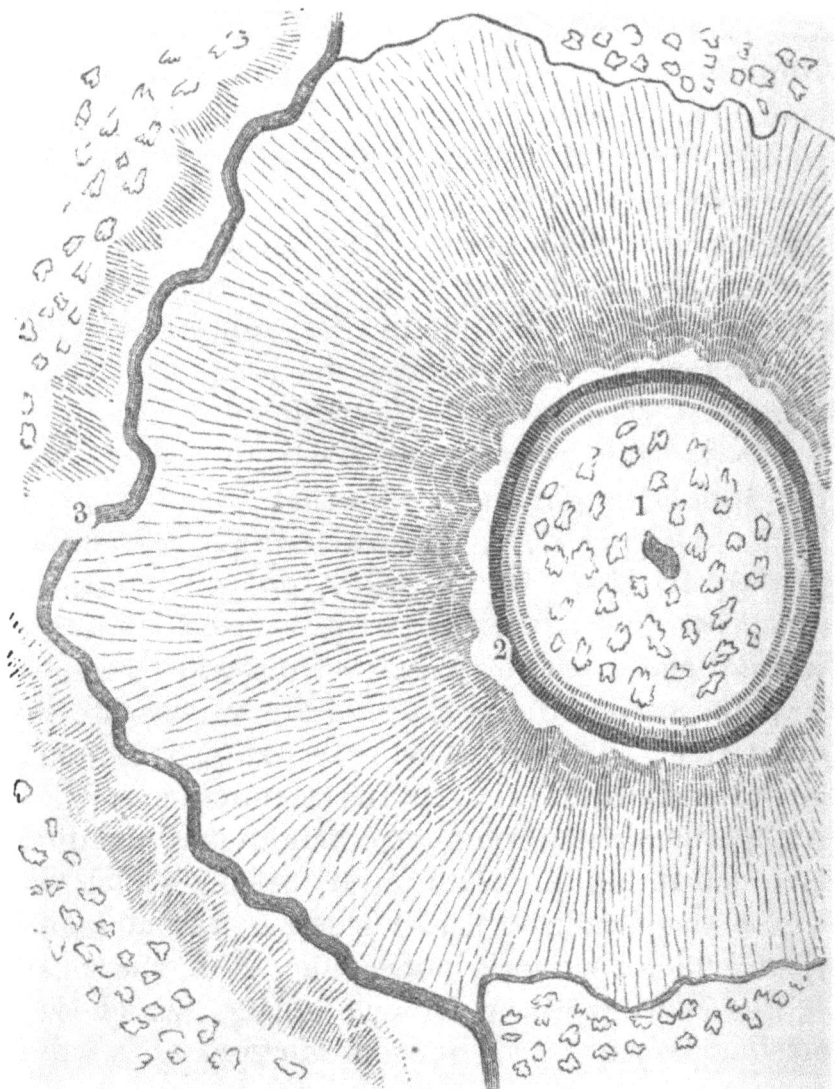

Fig. 6.—Great Fortification in the town of Red House. Figure 1, spring; 2, vallum; 3, Red House Creek.

Several miles below the Red House Creek, the remains of an ancient work can be traced. This is circular in form and about three hundred feet in diameter. Before it was disturbed the wall was four feet in height, with a ditch about six feet in lepth. Near the centre of this inclosure various relics have been found, many of which point to a civilization much higher than of those located in the Conewango valley.

In the year 1859, while exploring some tumuli in the vicinity of the Red House valley, we found numerous singular and interesting relics, among which were spear heads, six inches in length, with double barbs composed of masses of native copper; also several blocks of mica, which were in about the same condition as when chiseled from the granite of the Allegany Mountains.

It was near this valley where was found one of the most interesting relics ever discovered among the works of the ancient inhabitants. It was a flat piece of native copper, six inches in length by four in width, artistically wrought, with the form of an elephant represented in harness engraved upon it, and a sort of breast collar, with tugs on either side, which extended past the hips.

The great amount of copper implements and blocks of mica, that have been found, contradicts the theory of Mr. Squire, that the tumuli located in Western New York are not the work of the Mound-Builders. I am satisfied, beyond a doubt, that the Indian races never mined for mica or copper, neither did they bury either of these articles with the remains of their distinguished dead.

The Mound-Builders, for a long period of time, made the great chain of lakes a highway of travel and transported the copper, mined at Lake Superior, to the southern borders of Lake Erie, thence by land to the Cassadaga Lake and down the Creek to which it gives rise, to its junction with the Conewango.

An other important route was, by striking the Conewango, in the town of Dayton, and floating down that stream to the Allegany, and thence down that picturesque and winding river to its junction with the Ohio.

The circumstance of the Conewango and the Red House Valley being on and near the different routes to the southern rivers, may be the cause of the lavish distribution of copper in those sections.

Among the remarkable relics found in the vicinity of this last mentioned work, may be included an image chiseled from grey sandstone, which, for artistic design and elaborate workmanship, will compare favorably with modern art. It is given life-size and, no doubt, was designed to represent the form of head and face of the people then living. I have spent, with uncommon satisfaction, many hours in examining this wonderful specimen of ancient art. The head and features have a broad outline with high front. The brow is encircled with a plating of hair of fanciful ringlets that cluster round the intellectual forehead. Rough and primitive as is the school of art to which this image belongs, it carries unquestionable marks of a much higher civilization than any of the Indian races that have been discovered. Several archæologists who have examined this singular specimen, are of the opinion that it was an idol, one of the gods of their mythology, and worshipped in the mystical rights pertaining to their religion. As this ancient people worship the sun, it seems difficult to account for it on that hypothesis. But if it should prove to be an idol, we have a key to unlock one of the mysteries connected with a people once more numerous than the present population of America. A people, the sun of whose empire once arose beyond the northern lakes and extended south to where great rivers send down their turbid waters to meet the ocean's tides; and further still. to a land of wealth and flowers, where the golden fruits hang in tempting clusters, unborn of human toil, and thence o'er ocean isles, ere it is lost in the western wave.

Notwithstanding every climate on the American Continent has once been densely populated by a numerous people, their footprints are fast fading away. The last vestige of these wonderful inhabitants will soon be lost in oblivion, which has already cast its shadow across the " Acropolis of Ancient Athens and the Colosseum of imperial Rome. "

> " The long haired Greeks
> To him, upon the shores of the Hellespont,
> A mound shall heap; that those in aftertimes
> Who sail along the darksome sea shall say
> This is the monument of one long since
> Borne to his grave, by mighty Hector slain. "

CHAPTER III.

TUMULI IN THE TOWN OF COLDSPRING.

One of the most extensive sepulchral mounds bordering the Allegany River, in the State of New York, formerly stood on the west side of the river, in the town of Coldspring. This tumulus was located about five hundred feet from the river and some twenty feet above the highest flood. The location is grand and interesting. To the south and west is a broad level plain, embracing nearly two thousand acres of the most fertile land to be found in Cattaraugus County. This great interval is flanked on either side by precipitous hills which are cut asunder in but three directions; once by the great river that flows to the south, to the north and west by a branch of the Conewango and the Coldspring Creek.

For a description of this tumulus, as it appeared more than a hundred years ago, I am indebted to Gov. Blacksnake, the great Chief of the Six Nations. According to his estimate, it was about two hundred feet in circumference and twenty feet in height, and one of the largest trees he ever beheld was growing near its base, whose roots penetrated its side for fifty feet or more. When this mound was explored more than sixty years ago, it was found to contain a vast quantity of human bones. According to the recollection of the old Chief and that of his son (a very aged man who is now living on this Indian reservation) "cart loads" of bones were disclosed, many of which were in such a stage of decay that they crumbled to dust when exposed to the air. Great quantities of relics, such as gorgets, flint axes, arrow heads, and a great number of copper implements artistically wrought from masses of native copper which was brought from the mines of Lake Superior, were found with the bones.

This great tumulus is now nearly leveled down and the place where it once stood is a cultivated field. The earth of which it was composed is distributed over more than an acre of ground. So rich was this mound with decaying skeletons and relics of curious workmanship, that *now*, after more than sixty years have passed away, fragments of human bones, arrow heads and copper relics are found in large quantities at each successive plowing. In the spring of 1879, a few days after the ground had been plowed, in company with two boys we found fifteen arrow heads, a curious piece of copper and nearly a peck of fragments of human bones, all of which had been distributed in the soil during about fifty years of cultivation.

That this mound was very ancient is apparent from the condition of the remains being found in such a stage of decay, and from the size of trees found upon its summit, and adjacent to it. If the great tree referred to by the Indian Chief was near ten feet in diameter, as he supposed it to have been, that alone would reach back in the past nearly a thousand years, and it is not improbable that many generations of forests have

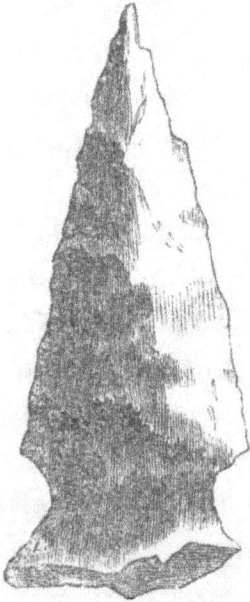

No. 7.—Shape of the arrow found in this mound.

grown and perished since that ancient tomb closed over scores of the dead.

At one of my interviews with Gov. Blacksnake, which lasted several hours, I derived much valuable information. This great Chief was a man in possession of vast native talents. He was familiar with the location of every prominent mound and fortification in Western New York, together with those found in Pennsylvania and the valley of the Ohio. In fact he was one of the most intelligent archæologists with whom I have come in contact. The subject under consideration at the time of this interview was the antiquity of the works of these ancient people, and whether they were in any way connected with the native Indians of America. In answer to the question, is there any alliance between the Indians and the Mound-Builders? he said,

" We have no knowledge of these ancient people, only from the mounds and fortifications that are very numerous in every section of country with which I am acquainted. We have no reliable traditions with regard to them, though we have traditions reaching back for a long period of time. "

Question. If a race of people as much developed in civilization as the Mound-Builders appear to have been, by their works which they have left behind, had lived in this country a thousand years ago, would not that fact have been transmitted to your generation?

Answer. "It is the custom with all Indian tribes to impart such information to their children, when very young, and if those ancient people were contemporary with the Mound-Builders, that fact would undoubtedly have passed through thousands of years. "

It was the opinion of the old Chief that the Indian tribes originated in America, and came in some mysterious manner long after the Mound-Builders had passed away.

Governor Blacksnake (as he was usually called by the white men) was the successor of Red Jacket, and the head Chief of the Six Nations known as the Iroquois Confederacy. He was in stature some six feet four inches in height and straight as an arrow. His skin was very dark, even for an Indian, hence his name Blacksnake, a name conferred upon him by the people of whom he was the head. I have known and conversed with many Indians, but have found none that for intelligence and native ability would compare with Blacksnake.

During our Revolutionary War he espoused the cause and fought with Washington and his men through that long and protracted struggle. For the energy and ability which he displayed, he was awarded a silver medal which was presented to him by the hand of Washington. In the latter part of his life he received many visits from the white people, and he would invariably haul from under his couch an old dilapidated trunk which contained his medal, and exhibit his proud memorial with a smile.

The last time I saw this distinguished Indian, whose name has gone down to history with that of Red Jacket, Cornplanter, and others, was at his home on the banks of the Allegany River. He had at that time braved the storms of one hundred and seventeen winters; but, old as he was, when lit up in conversation his countenance beamed with intelligence, especially when he was recounting the scenes of his early life and the time when he espoused the cause of American independence in the war of the Revolution, and led his braves into many a field of battle. To me it was a solemn scene to look upon that aged Chief, the last of the Iroquois, whose cycle in the history of that powerful people was nearly filled, and *he* the last great orb sinking in the western wave never to rise again. Blacksnake lived and died a Pagan and would favor no religion but that of his fathers. I was at that time informed by his son (who was our interpreter) that he had abandoned everything of a political character and was now floating around the boundaries of that invisible hunting ground, where he would soon reach

"Some safer world in depth of woods embrac'd,
Some happier island in the watery waste,
Where slaves once more their native land behold,
No fiend's torment, no Christian's thirst for gold. "

Blacksnake died at the age of one hundred and twenty-three years and was, at the time of his death, the oldest man in America. His remains are now reposing on the banks of his beloved river and near its rippling and stormy waters.

About two miles from the Allegany, up Cold-spring Creek, are two mounds, one of which has never been thoroughly explored. The other was leveled to the ground in the Fall of 1879. This tumulus is located on the farm of Wm. M. Brown, Esq., and was about thirty rods from the stream. It was originally nearly ten feet in height and about one hundred feet in circumference, and was partially opened several years ago, when human bones were found in the superincumbent earth.

At the time this mound was leveled, Mr. Brown generously contributed the use of three teams and several men accompanied with plows and scrapers. We commenced operations about 1 o'clock, p. m., and in about five hours reached the level of the surrounding soil. Here we found a quantity of bone dust mingled with coal and ashes. This was a very interesting discovery, as it shows that those ancient people practiced cremation to a certain extent at least.

The farm of Mr. Brown, on which this mound is located, is one of the most valuable in the county, on account of the richness of the soil and the grandeur of its surroundings. The valley where it is located, thousands of years ago, no doubt, contained a numerous population with cultivated fields and happy homes, rich with the luxuries which come from honest toil. But what a change! Sixty years ago that beautiful valley was a dense wilderness, thickly covered with tall pines that reached the height of two hundred feet or more. Who knows but that, in the far distant future, another great wilderness will be cleared away by a people who will look at the relics made by the present inhabitants with wonder and amazement?

ANCIENT WORK IN CARROLLTON.

Nearly opposite the village of Limestone, in the county of Cattaraugus, on lands of Job Moses, Esq., is a circular fortification about one thousand feet in circumference. This ancient work is located in the valley of the Tunaengwant, and is noted now as being a portion of the most extensive oil territory in America Sixty years ago this valley which extends in a southerly direction some fifteen miles or more, was a dense wilderness, a great pine forest, the home of the fallow deer and the black bear, whose right of possession was only occasionally disputed by the Indian while in search of food. Before the soil where this work is located was disturbed, it presented some remarkable and interesting characteristics. Some ten rods to the north is

a mountain stream that wends its way to the Tunaengwant that forms one of the most copious tributaries to the Allegany River. A large gateway
with breast works led towards this stream. In the
midst of this fortification once stood a number of
pine trees of giant size whose heads towered far
above the surrounding forest, several of which were
about twelve feet in circumference.

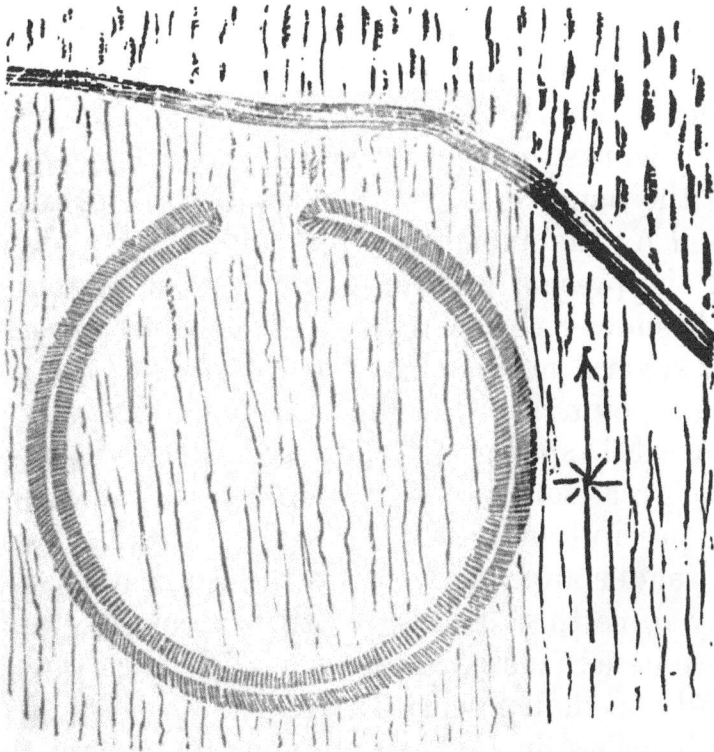

Fig. 8.—Ancient Fort near the village of Limestone, town of Carrollton.

Many of the stumps remain at the present time,
together with several others in the intrenchment
and on the parapet. About thirty years ago a
cellar for a house was dug within the boundaries
of the fortification, and human bones were disclosed

in great quantities, in such an advanced state of
decay that they offered but slight resistance to the
shovel and pick when brought to bear upon them.
Several years ago I had an interview with the man
who dug the cellar, and he informed me that the
bones were several feet in thickness as far as the
cellar extended.

Some writers are of the opinion that such circular fortifications were not designed for warlike purposes, especially those located in the valley; but
when we take into consideration that great quantities of human bones are generally found within
such works, it seems very apparent that they fell
by the hand of war. The mode of warfare practised by the ancient people cannot be understood
by the present inhabitants. We use cannon that
will shower the enemy with deadly missiles thrown
for several miles. They were confined to a rude
shaft pointed with a flint head and thrown by the
powerful spring of the bow. Notwithstanding
this great fortification is flanked to the westward
by precipitous hills, they are all too far removed
to render any assistance to the assailing party, for
any force they could give their arrows would have
become exhausted ere they could reach the outer
walls of that powerful rampart. Several hearths
and caches have been discovered within the intrenchment, and it is very likely that the besieged
party might have been imprisoned within the bas-
tion until starvation put an end to their existence,
or that they made a terrible charge from their ambush, vanquishing the enemy, and buried the dead

in a common sepulchre within the enclosure.

The valley of the Tunaengwant, together with that of Great Valley, Wright's Creek, and in fact nearly every stream that is bordered by fertile lands, were once occupied by the Mound-Builders. Vestiges of ancient works are numerous, such as military fortifications, sepulchral mounds, flint arrow heads, mica and copper relics, all of which point to the distant past, far beyond the Indian tribes.

As we pass up the Allegany, from its junction with the Tunaengwant, we find the faded remains of the works of the ancient people at every step. The ground now occupied by the village of Olean was, no doubt, once densely inhabited. The track of the Genesee Valley Canal passed directly through a sepulchral mound that disclosed human bones in the last stage of decomposition. This tumulus was oval in form and about forty feet in width by sixty in length and nearly ten feet in height.

ANCIENT WORKS NORTH OF THE ALLEGANY.

At the distance of about five miles, in a northerly direction from the village of Olean and near the Olean Creek, are the remains of an ancient fort that surrounds about three acres, and that is built in a circular form. This work shows unmistakable evidence of great antiquity; for more than fifty years ago, as I am informed by Joseph Palmer, Esq., of Cuba, pine trees, nearly four feet in diameter, were growing within the intrenchment.

Around this fort large quantities of relics peculiar to those lost and wonderful people have been found.

From this last mentioned fort we can trace them to the Genesee River and thence to Lake Erie and the southern shore of Lake Ontario.   In the town of Genesee Falls and about three miles north of the village occurs a tumulus that presents some wonderful and singular characteristics.   This work is nearly forty feet in height and covers more than an acre.   It is circular in form with a level surface on the top, which comprises about half of its diameter at the base.   Near the centre of this level surface is a well that extends below the base of the tumulus.

The object entertained by the builders of this extensive work in constructing a well which extended below the surrounding surface, where the Genesee River glides near its base, is one of the strange and wonderful mysteries connected with the works of the ancient inhabitants of America. I think it is not improbable that wooden structures were once placed upon this tumulus, which were occupied for defensive purposes in times of war, and that the well was designed to supply water to the warriors confined upon its summit.

From a Mr. Kingsley who has resided in the vicinity of this mound for nearly sixty years, I have gathered much valuable information.   He informs me that that part of the tumulus facing the river has been partly explored, and human bones have been found, to use his words, " in quantities sufficient to load a dozen wagons. "

This work carries the marks of great age; trees several hundred years old are now growing upon its surface. The celebrated Indian Chief, Sangeuwa, whose death occurred about fifty years ago, at the age of more than a hundred years, according to his reckoning of it, said that this tumulus was not the work of the modern Indians, but was built by an unknown people who occupied the country far beyond any Indian traditions. Relics of curious workmanship have been found in great quantities within the mound and adjacent to it. To the westward of this tumulus and at no inconsiderable distance is a hill called Fort Hill, thus named from its containing an ancient fortification which is now well defined, notwithstanding it has, no doubt, been exposed to the frosts of more than a thousand winters.

As we pass in a northerly direction from the alluvial borders of the Genesee River, we can trace these ancient people over precipitous hills and along the winding streams whose waters seek a repose in the placid bosom of Erie or Ontario.

In the town of Aurora, about thirteen miles in a southerly direction from the city of Buffalo, are several ponds located about a mile north of the village. Near these pools of pure water are a number of mounds varying much in size, which, when explored, were found to contain implements of copper, mica, and other relics peculiar to the mounds located in the valley of the Ohio.

The ground occupied by the village of Aurora

is supposed to have once been the theatre of a terrible conflict, as vast quantities of spear and arrow heads are found in every direction. In several cellars that have been dug within the village, human bones have been disclosed in great quantities.

In Turner's History of the Holland Purchase, mention is made of a number of mounds and fortifications, which have been mostly obliterated. Among them is one of large dimensions, located in the town of Cambria. These works were examined by Mr. Turner, in the year 1823, at which time they were well defined. The fortification was located on elevated ground from which could be seen the glistening waters of Lake Ontario, and the large area of fertile land bordering its southern shore. The ditch and parapet enclosed about six acres, in the centre of which was a pit about five feet deep from which was taken about one hundred human skeletons, all of which were promiscuously buried in a heap together; they consisted of persons of both sexes. Many of the skeletons were those of aged persons whose teeth had fallen out, and the alveolar processes had been absorbed before they fell victims to the arrows of the enemy. Barbs and arrows were found in great numbers. The finding of so many human remains within the intrenchment points to the termination of the battle in favor of the assaulting party. I have no doubt that those vanquished heroes were driven with savage fury towards the border of the great lake, and were there forced to make a last stand to vindicate a cause and save their wives and children from the dreadful spring of an enemy mad with revenge and insane with the terrible frenzy of superstition.

Near the village of Leroy, at the time of its settlement by the whites, was a large fortification with a deep ditch. This work was circular in form with several gateways leading into the enclosure. Mr. Schoolcraft, author of the "Indian tribes of America," says, "it was a powerful bulwark of defence." The weapons of warfare peculiar to those warriors have been found; also stone axes and pipes made of clay with bowls in the form of the human head. Numerous forest trees of great size were growing upon its surface, whose roots penetrated the bones of the slumbering dead.

In the town of Lewiston were formerly two large mounds, each of which contained human remains. Copper relics together with various flint implements wrought with mechanical skill were exhumed.

On Buffalo Creek several mounds have been explored; also at Tonawanda, and in fact in almost every section bordering the lakes and tributary streams.

From Lakes Ontario and Erie we can readily follow the trail of the ancient warriors to the Genesee River and thence along fertile valleys to the Allegany, where they could launch their light canoes and gently move with its current to more congenial climes.

That many of the relics which have been found betwixt the Allegany River and the southern shores of Lakes Erie and Ontario were used by the mod-

ern Indians, is very probable. Large numbers of
warriors traversed the different routes leading to
the Allegany but a short time previous to the battle of Braddock's Field. Numerous Indians from
Canada, who had allied themselves with the French,
were engaged in that battle and availed themselves
of the most conspicuous route to reach the head
waters of the great river; and then the Iroquois
held undisputed sway for a long period over the
territory south of Lakes Erie and Ontario. That
the mounds in Western New York, which show
such evidence of great antiquity, were built by a
people whose religion and mode of warfare were
similar to the builders of the great mounds and
fortifications in the Ohio and the Mississippi Valley, is very apparent. Numerous relics found in
Western New York point to the builders of the
mounds as worshippers of the sun. Mica, as heretofore suggested, was connected in some mysterious
manner with their mode of worship. Engravings
on stone, representing the sun and the heavenly
bodies, are frequently found. The large rock located on the Allegany River and known as the
"Indian god," to which reference will be given in
another chapter, bears witness that the great central luminary was the prominent object of worship
with the people who constructed the mounds, not
only in Western New York, but in most of the
fertile valleys located in the western and southern
states.

CHAPTER IV.

ANCIENT WORKS IN CHAUTAUQUA COUNTY.

Chautauqua County, the most western in the State of New York, is bounded on the east by Cattaraugus and on the south and west by Pennsylvania, and on the north its boundary line contends with the billows of Lake Erie for more than thirty miles. Vestiges of works of the ancient people are numerous in the County of Chautauqua, many of which are found near the shore of the lake.

In the town of Portland a number of roadways have been discovered, underlayed with stone and covered with sand and gravel to the depth of several feet. The object of building such graded roads has been a mystery, especially to those who suppose the builders were allied to the Indian tribes.

Not far from the eastern boundary of the village of Fredonia and near the Canadaway Creek, is an ancient work that attracted much attention among the early settlers of the town of Pomfret. This fortification is located on a precipitous hill whose top is nearly level, and extends to either bank across its level summit. Nearly in front of this breastwork was a deep pit built, no doubt, for storing provisions and implements of war. I was informed more than thirty years ago by one of the

early white settlers that a great amount of arrow
heads and spear heads, some of which were composed of copper, together with a quantity of charred corn was found in the pit.   Adjacent to this
intrenchment fragments of human bones have been
disclosed in such an advanced stage of decay that
they whitened the soil in which they were laid.

The location of this work was well selected, for
it commands a view of a large amount of territory
bordering the lake, and from its elevated position
its warriors could readily detect the approach of
an enemy from several directions and shower their
deadly missiles upon the heads of an invading
army.

In the town of Sheridan, near the present location of the New York and Erie Railway, was formerly an extensive fortification surrounding several
acres of land.   This once prominent work was
constructed with remarkable care.   Its position is
more than two hundred feet above the waters of
Lake Erie and where its sentinels could overlook
the surrounding country for many miles, behold
the bosom of the placid lake painted in purple and
gold by the lingering rays of the setting sun, and
hear the rippling waters of Walnut Creek, since
rendered immortal by its fertile valley producing
a tree the most extraordinary in size east of the
Rocky Mountains.   Evidences of ancient cultivation in the vicinity of this work, for some distance
in either direction, is well defined.   Numerous
mortars, pestles, arrows and various other implements have been found, together with several ex-

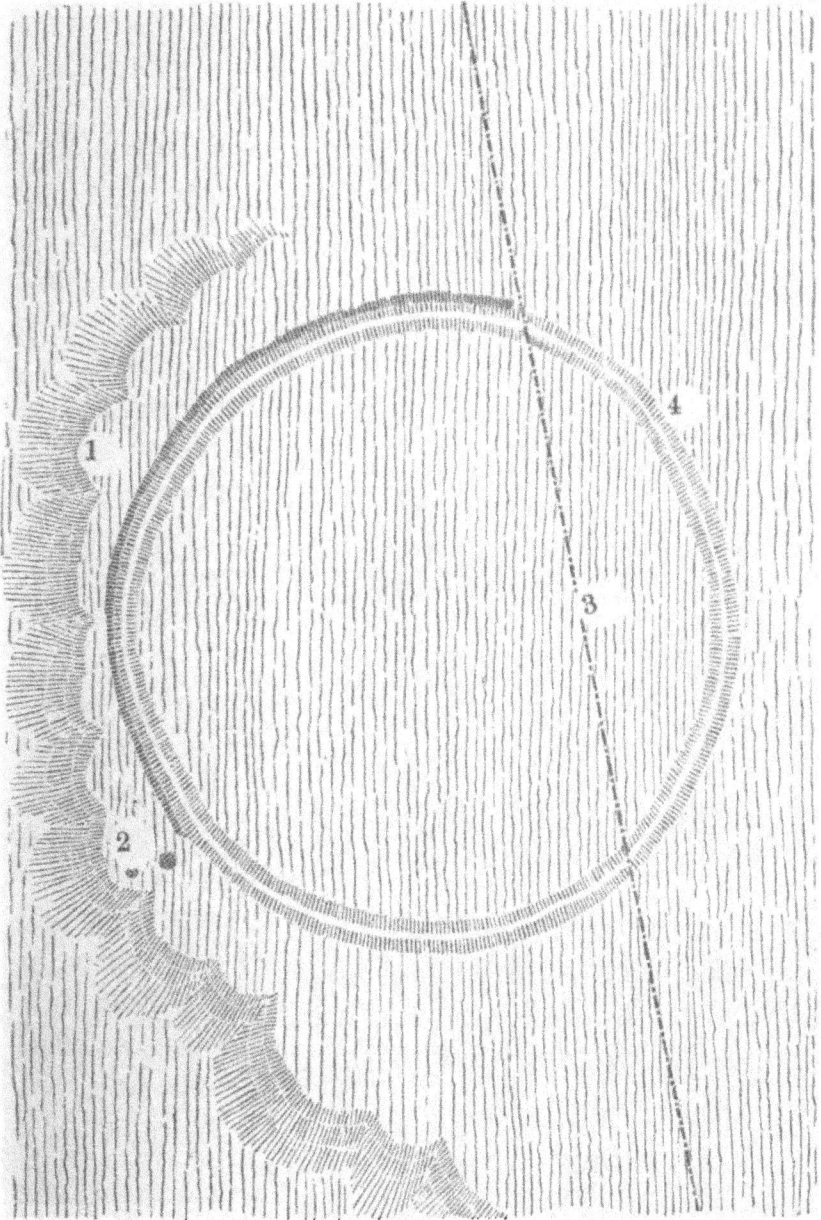

Fig. 9.—Circular fortification in the town of Sheridan. Fig. 1, steep bluff; 2, hearths; 3, road; 4, intrenchment and parapet.

tensive pits which were nearly filled up by the slow accumulation of decayed substances that have been

going on for hundreds of years.   As is always the
case, human bones have been found in large quantities buried within the intrenchment.   In the summer of 1870 a grave was opened that contained a
great number of human skeletons, comprising persons of both sexes and of all ages.   These bones
were promiscuously buried together in a common
grave.

The circumstance of finding the skeletons of
male and female, and of youth and old age mingled
together would go to show that the builders of these
extensive fortifications and their opposing foes
practised the savage and murdered the father,
mother and child, who were driven from house and
home and who sought protection within the walls
of the fortification.

As we travel southward, the distance of about
six miles, we strike the northern shore of two beautiful lakes whose surplus waters join the Conewango, a sluggish stream that enters the Allegany at
Warren, in the State of Pennsylvania, and then
flows on through the Ohio and Mississippi Rivers
to the Gulf of Mexico.   This sheet of water is
more than six hundred feet above Lake Erie and
but a few hundred rods south of the summit that
divides the waters betwixt the great lake and the
Gulf of Mexico.

WORKS AT CASSADAGA LAKE.

Cassadaga Lake, so named by the Indians, is
surrounded with scenery grand and interesting.
The beautiful village of Cassadaga is located

near its waters. To the east and west it is flanked by gently sloping hills dotted with elegant dwellings which are surrounded with many of the most valuable farms in Western New York.

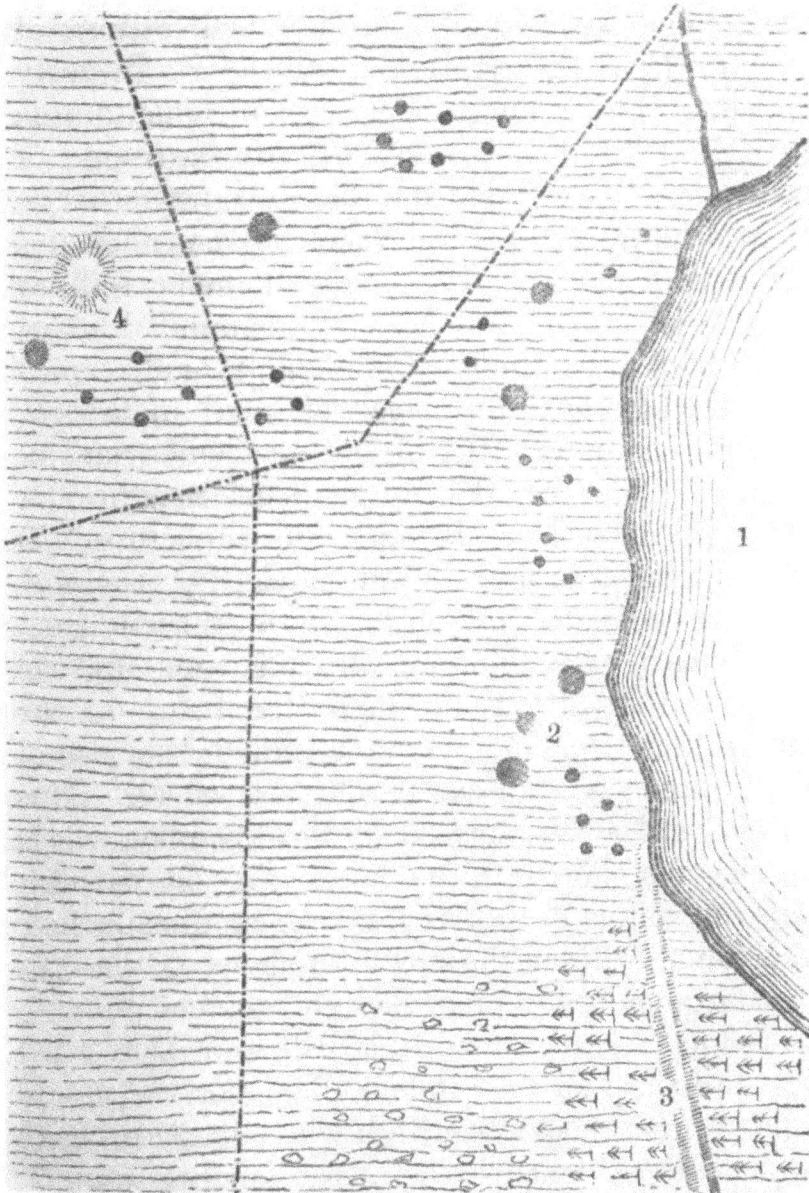

Fig. 10.—1, Cassadaga Lake; 2, hearths and caches; 3, graded way; 4, mound.

Some of the most extensive ancient works ever
discovered in Chautauqua County were located
here.   The mounds, breastworks and graded roads
in the vicinity of these lakes were thoroughly surveyed, some years ago, by Hon. Obed Edson, of
Sinclairville, his description of which was formerly
published in "Young's History of Chautauqua
County."   Mr. Edson says, "At the extremity of
the cape which extends from the south-western side
far into the lower of these lakes, is a curious and
conspicuous mound.   Its longest diameter is about
seven rods; its shortest five.   Its summit is about
twelve feet above the level of the lake and is about
eight feet above the low neck of land in its rear
that connects it with the higher and wider part of
the cape.   Whether it is an artificial structure, or
the work of nature, is open to conjecture.   It
seems however to have been anciently occupied,
for the usual relics have been found there in great
abundance.   Stretching across this cape for the
distance of twenty rods along the brink of the
plateau that rises about twelve rods in the rear of
this tumulus, was an earthen breast work.   Thus
were several acres inclosed by these earthen works
and the two shores of the lake.   In the vicinity,
large quantities of pottery and stone utensils have
been found.   Near the northern shore of the lake
was a large mound which, though frequent plowing
has reduced its dimensions, is still four or five feet
high and three or four rods in diameter.   It is said
to have been twelve feet high when first seen, with
forest trees of centuries' growth standing upon it.

About 1822 this mound was excavated, and a large number of human skeletons exhumed. · Extending from an extensive fire bed in the neighborhood of this mound, in a north-westerly direction, a distance of sixty rods or more on the east side of the lake, was an elevated strip of land, the width of an ordinary turnpike, bearing the appearance of having been once a graded way. The traces of this ancient road are plainly visible. At various other places around Cassadaga and along the shore of the lake were numerous caches and extensive fire beds, or hearths, with an abundance of coal and ashes buried deep in the ground. Skeletons have been exhumed in many places, and arrows, pottery and stone implements in great profusion. "

As I have heretofore suggested, one of the most important routes from the southern shore of Erie to the rich and picturesque valleys of the Ohio and Mississippi, was by way of the Cassadaga. The lakes above referred to comprise one of the head waters; and at the present time the stream fed by the lakes can be navigated with canoes; and from a few miles to the south rafts of lumber are floated to the southern cities.

The place where the village of Cassadaga is located was once a place of great importance to the ancient people. Here they had the first great depot after landing in the harbor of Dunkirk. In idea I can imagine that thousands of years ago much of the riches gathered at Lake Superior were transported along this route on their way to the distant south, and that hundreds of travelers loaded with copper and other treasures were daily moving up a winding stream on their way to this important summit.

It is stated in a paper written by Dr. Cheney, in 1859, that the skeletons found in the mounds at Cassadaga were those of giants, and that one in particular measured seven feet and five inches. I suppose he got that information from some persons who saw them at the time they were exhumed, and their organ of marvelousness greatly exalted. That the Mound-Builders were a trifle larger than the present type, is very probable; but that they were giants eight and ten feet high is all fabulous. I have seen many skeletons from mounds in different states, but have seen none that will much exceed the present people now living. At the Centennial, in one of the annex buildings, was a large amount of fragments of skeletons from the ancient tombs in West Virginia, Missouri, Ohio, and the Mississippi valley, and I saw none that would exceed the Indian tribes of America.

The subject under consideration has enough of the marvelous about it to gratify almost any imagination without resorting to giants. When we look over the Continent of America and find a race of people who once held empire over this broad domain, now slowly and surely passing away; and then into the midnight of the past and find such unmistakable traces of a once numerous and powerful people now lost in oblivion; and then again that we, as a people, may be standing on the brink of ruin with religious and political leaders lost to every principle of honor, and slowly undermining the glorious fabric of American greatness.

Who knows but in the distant future America with its colossal structures will be in ruins, when we consider the warlike spirit of man and the strange and dreadful convulsions of nature, that carry in their train vast and sudden destruction? "What the headlong swoop of armies has spared the minutest things of earth have claimed for their own, for that wonderful agency has partitioned the depths of the ocean with coral reefs, and eaten away the well chiseled marble that once bore witness to the pride of human greatness." I am aware that we, as a people, believe that we are crowned with everlasting safety, on account of our republican institutions, our liberty of speech, and our great range of religious and political freedom. The ruined cities of the ancient world no doubt thought so once; but "hearken to the voice of history." Look at Babylon, the praise of the whole earth, with its magnificent streets and fertile gardens irrigated with crystal streams to give health and beauty to its inhabitants; adorned with magnificent temples that, for enchanting beauty, bid defiance to the most skilful artists of this nineteenth century. But "Babylon has fallen," treachery, superstition, war and murder "brought the hectic flush that corroded its inward frame, and a blaze that gave place to blackened ruins and smouldering ashes."

"How nations sink by darling schemes opprest,
When vengeance listens to the fool's request,

Fate wings with every wish the afflictive dart,
Each gift of nature and each grace of art ;
With fatal heat impetuous courage glows,
With fatal sweetness elocution flows.
Impeachment stops the speaker's powerful breath,
And restless fire precipitates in death. ''

WORKS AT SINCLAIRVILLE.

At the distance of six miles in a southerly direction from the ancient ruins at Cassadaga, and where the village of Sinclairville now stands, were formerly some very prominent and interesting works. This location was well adapted by nature for defence against a hostile army. On the east it is flanked by high hills, and on the north a rapid stream dashes down through a deep gorge flanked to the east and west by precipitous bluffs. In a westerly direction is a prominent hill on whose summit has been found a large granite bowlder in which is a deep cavity formed by human hands. This interesting relic was very likely designed for preparing corn and grain for domestic purposes.

Why those ancient warriors should have selected the summit of a high hill as the place to prepare and grind their corn, is difficult to understand. But, it is very likely, it was on account of its elevated position from which they could command a view for many miles of the broad and magnificent valley of the Cassadaga, together with the high hills in the town of Ellery, which separate the great valley from the waters of Lake Chautauqua. In a southerly direction from the village, on either side of Mill Creek, is a broad valley that contains many valuable farms noted for their uncommon fertility, especially for the production of Indian corn. This valley follows the course of the stream to its union with the Cassadaga, where it distributes its pure and sparkling waters drawn from living springs.

When the land where the village of Sinclairville is located began to be explored by the white man, he was astonished to find an ancient battle field, where were fortifications, breastworks and various implements designed for warlike purposes peculiar to a pre-historic age. These works have been thoroughly examined by Hon. Obed Edson, a resident of the village. Though they have given place to churches, stores and dwellings of various kinds, a correct record has been preserved.

In reference to those works Mr. Edson says in his historical address, " At the first settlement of Sinclairville an earthen breastwork extended westerly from a point near the stone blacksmith shop and Baptist Church for a distance of about forty-five rods parallel to the steep bank at the northerly boundary of the plateau upon which the central portion of the village is situated, and distant from the bank about two and one-half rods. Near this earthwork and where the garden of the village tavern is situated, buried beneath the surface of the ground were large quantities of stone that had been charred and broken by fire. Even to this day arrow heads of flint, stone hatchets and other rude instruments of the stone age are found. Between this work and the steep bank of the hill the first highway of the village once ran. Running for many rods easterly and westerly, near to and parallel to the southern boundary of this plateau, was a similar breastwork. Situated between these embankments seemed to be the principal fortification. It was an extensive circular earthwork having a trench upon the outside and near the small rivulet that runs through the village. Along its southern side was a gateway. This little brook that crosses Main Street ran through the northern portion of this fortification. It inclosed six or seven acres of what is now a central portion of the village. A part of Main Street, portions of other streets and the village green, all were included within its boundaries. Its earthen walls, where they crossed Main Street nine or ten rods west of the village green, were about four feet high, and required some labor to level them when that street was first constructed through the embankment. Upon the high bluff to the west, that rises precipitously from Mill Creek, was another circular earthwork within which was a deep excavation. ''

The ground where Sinclairville is located would seem well selected by those ancient warriors to construct a powerful rampart of defence against the approach of a hostile enemy. The earthen breastwork facing the south would baffle the designs of the foe; and, if properly manned with veteran warriors skillful with the bow, they could defy the arrows of a powerful army. If the breastwork should fall into the hands of the enemy, they could fly through the gateway and find protection within the walls of their circular fortification. Again, if they were threatened from the north, those in the fortification located upon the bluff could stagger the offensive army whilst attempting to pass the gorge.

At the distance of about one mile south of the village of Sinclairville, in the town of Gerry, was formerly an extensive earthwork, and near it an ancient cemetery where the bones of nearly one hundred human beings were exhumed. About sixty years ago, when this burial ground was first explored, fifty or more human skeletons were removed, and some thirty years subsequent to that time, on one occasion, twenty-five more were disclosed. The ancient fortifications and repositories of the dead make it very apparent that the whole valley of the Cassadaga and its surrounding country was long ages ago populated by a warlike and numerous people, and that the valuable farms now displaying great herds of cattle in its luxuriant pastures and fertile hills, were cultivated fields.

South of this last mentioned work and at the distance of about one mile was formerly another fortification oval in form and enclosing about one acre. The circumstance of two such military works being placed in such close proximity, and both located on a level plane, would indicate that they were constructed by the two contending armies that were arrayed against each other; and that the crumbling skeletons found in that ancient grave yard would point to the horrors of war and the carnage of battle.

Further down the valley and on either side of a

small stream are several ancient ash heaps produced, very likely, by extensive and long continued fires designed for giving alarm of danger and for preparing food for the army. It however is not improbable that they lived to some extent in communities, as has been the case with various pagan nations.

At the distance of three-fourths of a mile from Bucklin's Corners, upon elevated ground bordering the valley of the Cassadaga, was once an extensive fortification oval in form and enclosing about seven acres. Notwithstanding the ground is cleared and cultivated, a large part of the work is now visible. In selecting this locality for protection and defence these ancient warriors prepared themselves with water. Four springs, even at the present time, pour out a copious stream that forms a rivulet joining the turbid waters of the Cassadaga.

At the village of Bucklin's Corners was an ancient cemetery where a great number of human skeletons have been exhumed. These were, no doubt, the bones of those who fell in battle, for they are found very near the surface of the ground.

At the distance of about a mile across the valley of the Cassadaga and near sixty rods from the line that divides the town of Gerry from that of the town of Ellery, are some interesting and prominent fortifications. These, like most of the ancient works found in Western New York, display a military significance; they are situated on the farm of John A. Almy and near the terminus of a gently sloping hill whose base is about fifty feet down a precipitous bluff. At the time these works were constructed this sloping hill was much more extensive than at the present time; but it has been gradually washed away by the waters of a rapid stream that have dashed against it for thousands of years. This brook contains the purest of water, for it is drawn from living springs that flow from near the summit which divides the great valley of the Cassadaga from that of Chautauqua Lake.

The ground where these works are located presents a beautiful appearance, for they are surrounded by a grove of second growth maples whose healthy condition and remarkable luxuriance tell

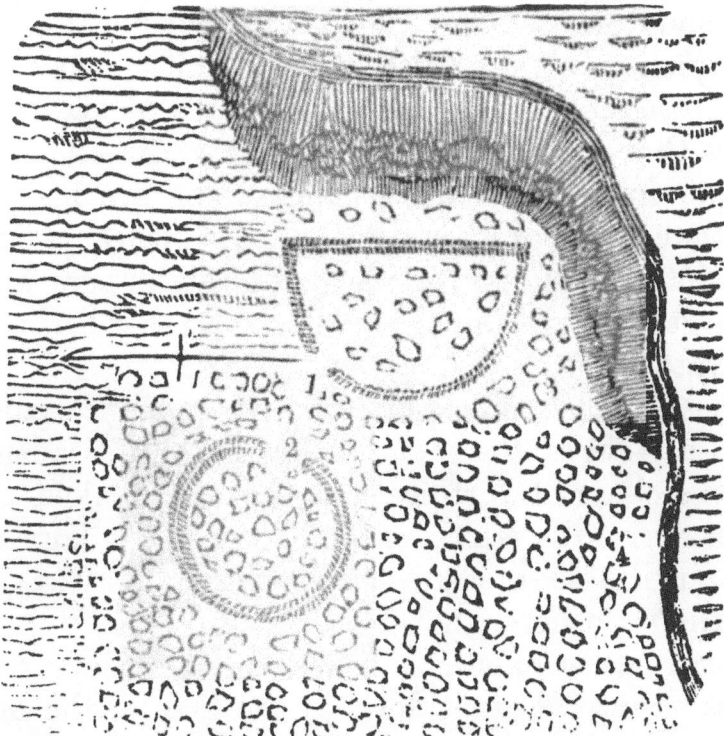

Fig. 11.—Fortifications in the town of Gerry. Figs. 1 and 2, gateways; 3, bluffs; 4, mountain stream.

of an ancient cultivation. Two enclosures are located here, each of which consists of a ditch and
parapet; and, notwithstanding the many centuries
that have elapsed since their construction, they now
are well defined. The larger of these fortifications is in the shape of the letter D with the straight
line fronting the eastern border of the bluff. The
ditch surrounding this work was originally some
eight or ten feet in depth, and when thrown out
and properly placed, would form a well at least
eight feet in height.

Near a gateway located on the southern side of
this fortification is a ravine or graded way which
leads directly down this rugged bluff, until it meets
the mountain stream. This was constructed to
supply the warriors with water, whilst awaiting
some terrible battle. At the distance of six rods
across this graded way, in a westerly direction, is
another fortification whose dimensions are less in
size than those of the former. This work is circular in form and surrounds about half an acre, and
has a gateway facing the graded road that leads to
the water.

These works being placed on an eminence command a view of the great valley of the Cassadaga
and the chain of fortifications that are flanked by
the eastern hills which extend from the village of
Sinclairville to the ancient cemetery located at
Bucklin's Corners.

About sixty rods in a southerly direction, and
across the stream that flows at the base of the hill
upon which the above mentioned works are loca-

ted, was formerly another fortification enclosing about three acres. This work, like the larger of the other, was a true circle facing the south and divided in the centre by a straight line. It is located on a level plain that forms a portion of the western boundary of the valley of the Cassadaga. From the western side of this fortification was an extensive ditch, about sixty rods in length, which reached the bed of the stream at a sufficient height to force its waters within the enclosure. This ditch was undoubtedly constructed for the purpose of supplying water for the use of the soldiers during a protracted battle. In the summer of the year 1869, in company with J. L. Bugbee, Esq., of Stockton, a gentleman of intelligence and one who has made a life study of this important and interesting subject, archæology, we made thorough investigation of these works. Aside from our examinations, we gathered much valuable information from Mr. S. Morgan Tower, the owner of the land on which the last named fortification is located. Within this enclosure was formerly a large number of hearths, caches, flint-arrow heads, ancient pottery, and other implements used in war by that unknown people whose history is buried in the great ocean of time.

This work which has stood for ages in a dark and unbroken wilderness, and on whose surface numerous generations of forests have been born and have perished, is now a cultivated field; naught now remains of its earthen walls; fragments of pottery and other relics are all that remain to tell of the existence of that ancient field of blood.

Around these works a vast quantity of relics have been turned up by the plow. Mr. Tower has in his possession several of the most interesting of any that I have discovered in this section of country. One in particular was a gorget composed of a kind of stone which is very hard and resembles a clouded cornelian, a stone not indigenous to this country. This relic is flat on one side and oval on the other, being larger in the centre than at either end. The ends are both perforated in such a manner as to receive a string with which it could be suspended from the neck. This relic is about seven inches in length and wrought with remarkable skill. Every part is perfect and highly polished. The great amount of labor that has been expended upon this gorget makes it very probable that it was worn by some distinguished personage who was connected with those warlike people. Whether it was worn by a priest whose faith and prayers had been fully vindicated amid the monuments of death, or some mailed warrior who had led his braves to some field of battle and beaten back the enemy with dreadful carnage, is difficult to determine.

With regard to the antiquity of these works, there has been much speculation. Some suppose they are very recent, when compared with the great works in the Ohio and the Mississippi valley. Whether this is the case or not, they were constructed long ages ago.

Within the larger intrenchment located on the

farm of Mr. Almy, was a pine tree that attracted the attention of the early settlers of the country on account of its great size. Located as it was upon an eminence, and towering towards the clouds to the height of more than two hundred feet, it was justly called the monarch of the forest. This tree was planted directly in the midst of the ditch and near the gateway leading to the graded road that extended to the stream. About fifteen years ago it was cut by Mr. Almy and when measured, as it lay upon the ground, was upwards of two hundred feet in length. The stump now remains and is nearly six feet in diameter, and when examined by myself and Mr. Bugbee, we readily counted eight hundred annual rings, leaving a large number on the outer side that were obliterated by decay.

Determining the age of trees growing upon these works does not determine the time of their existence. Other trees may have grown and gone to decay long before this monarch raised its head above the walls of this ancient fortification. The cause of this tree attaining such a great size, and even of those now surrounding the above works and presenting uncommon vigor, was the ancient cultivation and some elements of fertility mingled with the soil.

That the valley of the Cassadaga was occupied by a large population hundreds of years before the solitude was broken by the hand of civilization, is very apparent from the great amount of ancient works found here. Traces of an ancient cultivation of the soil are found upon every hill top and

extend across the summit to the southern shore of
Lake Erie. The bark canoes and rude vessels that
transported the copper from the mines of Lake
Superior have given place to merchant ships whose
snow-white wings beckon the coming of the gale;
and magnificent steamers plow their waters undis-
mayed by the music of the blast. In this great valley that formed such a prominent route to the
southern rivers, and where in ancient days fortifications were erected and battles fought, can now
be seen beautiful dwellings and elegant farms.
That dark wilderness which seventy years ago was
covered with a mantle of living green, has been
made to " bloom and blossom as the rose. " On
ground where once were held nocturnal revels at
some theatre of superstition can now be seen the
uplifted arms of the Christian Cross. Civilization
in its rapid march has entered the valley that ages
ago was changed from a garden to a desert. The
dark waters of the Cassadaga, on whose surface
once glided the rude craft of the ancient warrior,
is now checked in its onward march, and its powerful arm made a willing slave to the majesty of
civilization. The ancient cemeteries, where mourners shed the parting tear, have given up their dead
and a higher race has built upon their ruins.

VESTIGES OF WORKS ON CHAUTAUQUA LAKE.

On the eastern shore of Chautauqua Lake, about
one hundred rods from the Narrows, and separated
by the distance of about five rods occur two
mounds whose diameters are about equal, being

respectively sixty-six feet with an altitude of six feet.

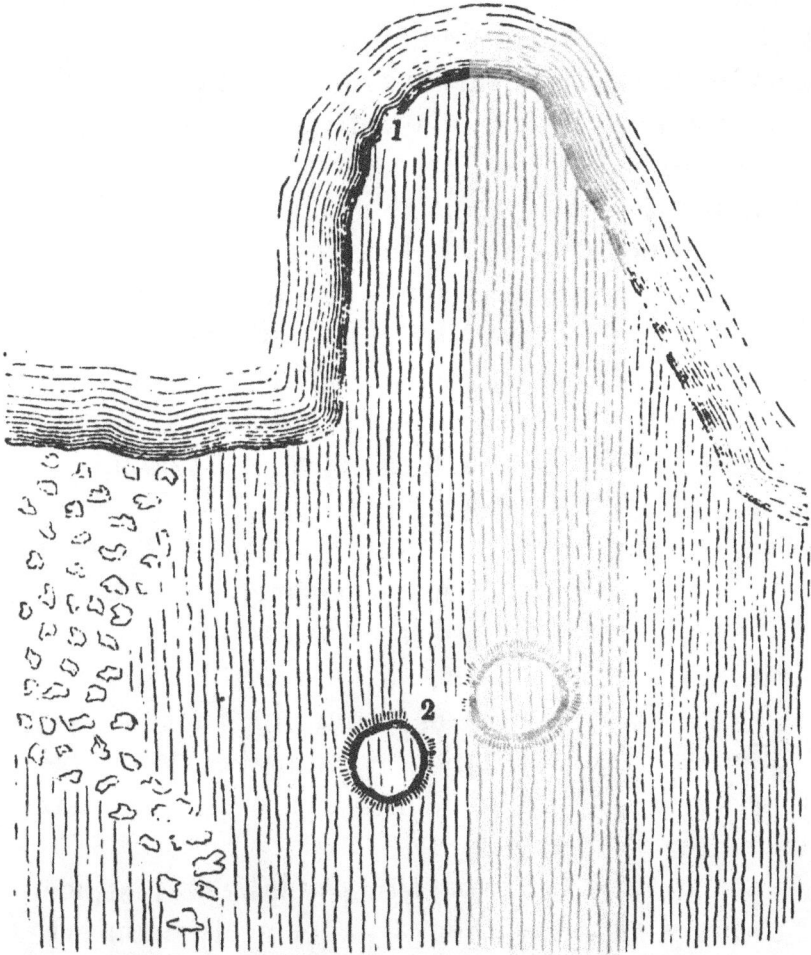

Fig. 12.—Chautauqua Lake. 1, eastern shore of the lake at the narrows; 2, tumuli.

Near these mounds glides a crystal stream whose murmuring waters mingle with those of the placid lake. From the appearance of the soil in the vicinity of these tumuli, and on account of its fertility, it was undoubtedly cultivated to a great extent to supply that ancient people with corn and other vegetables, whilst the lake supplied them with fish. I can imagine a time in the long ago when that beautiful lake was dotted with the light canoes of that long lost people hauling in their lines and dancing upon its stormy waves, whilst its borders were eloquent with a busy throng cultivating the soil, teaching their children the arts of war and the sacred mysteries of their religion.

At no inconsiderable distance from the above mentioned tumuli was once an ancient cemetery consecrated to the dead. This ground, once so sacred to the ancient people, is now occupied by the village of Dewitville. Human remains which were no doubt those of fathers, sons, husbands and wives, entombed amidst the tears of those whom they loved and expected to meet again "beyond the river of death," were mingled with the dust.

Nearly fifty years ago, as I am informed by one of the early white settlers of Dewitville, vast quantities of relics have been found in that vicinity. Whilst digging a cellar he exhumed about a peck of arrow heads, all of which were at least four feet below the surface.

Not far from the outlet of the lake and adjacent to the city of Jamestown was formerly disinterred the remains of a Mastodon whose bones were in a far better state of preservation than those of the ancient man found in the vicinity of Dewitville.

Chautauqua Lake located about seven hundred feet above Lake Erie and surrounded by the most beautiful and interesting scenery and composed of water drawn from subterranean fountains, makes it one of the most charming sheets of water that ever glistened in the sunlight. So great is the interest taken in this wonderful lake, that during the summer months thousands of people are drawn to its shores, and numerous steamboats loaded with living freight continually ply upon its waters.

If the people who have arrived at the highest civilization yet attained by man, are charmed with Chautauqua Lake and the beautiful scenery with which it is surrounded, it is no wonder that ancient man was attracted to its pure and sparkling waters, though it was buried in the dismal forest.

From the appearance of an ancient cultivation, in places between Lakes Chautauqua and Erie, I am led to believe this was another one of the important routes from the great lakes to the Allegany River. By striking the lake in the vicinity of Mayville they could pass down the outlet to its junction with the Cassadaga, and thence to the Allegany River near the village of Warren. The great amount of relics found in the town of Ellery and the borders of the lake, on either side, and the appearance of the soil when first cleared of the forest by the white man would indicate that far away in the past this section was densely populated, and that the soil was cultivated to a large extent. The circumstance of finding charred corn in several hearths found in this section, would indicate that these ancient people cultivated maize. As the land surrounding Chautauqua Lake is well adapted to the growth of corn and placed amidst the palaces of nature, it is not improbable that villages were once erected there, which, millenniums ago, gave place to a dark and dismal forest that changed the habitation of man to that of wild beasts.

On ground now occupied by the village of Fentonville the footprints of ancient man have been discovered, and at no inconsiderable distance from the village human bones have been found in great abundance, which would indicate that an ancient cemetery was once located here. This section, like the valley of the Conewango and that of the Cassadaga, was originally covered with a magnificent forest of pines, many of which had been growing for centuries at the time Fernando Cortez conquered the Montezumas. Several years ago, under the roots of an aged pine that had been leveled by the wind, were discovered the osseous remains of men who, from their appearance, had been reposing in the ground long before the aged tree began its life.

The village of Fentonville is located on the Conewango and but a few miles below where it joins the Cassadaga. Surrounding the village is a broad interval of land that contains a soil remarkable for its fertility; and maize which constituted the vegetable food of those ancient people grows in great luxuriance.

The circumstance of the adaptation of the soil to the development of their most prominent food, and the large amount of human bones which have been disclosed, would indicate that a large population of an ancient people occupied that rich and beautiful valley long before the wilderness of majestic pines were ushered into life.

As the junction of the Conewango with the Allegany River is but a few miles in a southerly direction from the above mentioned village, it is very apparent that a numerous people for a long and unknown period occupied every available plateau of land; and that the majestic hills that flank the Conewango on either side resounded with the voices of the hunters.

CHAPTER V.

WERE THE ANCIENT WORKS FOUND IN WESTERN NEW
YORK CONSTRUCTED BY THE MODERN INDIANS OR
BY THE MOUND-BUILDERS?—REMINISCENCES OF THE
IROQUOIS.

The fact that Mr. Squier, after making some personal examinations of a few of the works in Western New York, has pronounced them the works of
Indians, has had great weight with modern investigators who have adopted that theory.

When this Continent was settled by the white
man it was everywhere occupied by native Indians
whose implements of warfare were peculiar to the
stone age, and great quantities of relics were found
in every section. In New England, which then
contained a powerful tribe of Indians, mounds and
fortifications similar to those found in Western
New York have never been discovered. The Iroquois, after they had been united in a league with
other tribes, rapidly increased in numbers and
acquired an intemperate thirst for military glory.
If we can believe their traditions, by their military
bravery they attained an empire which extended
from New England to the northern lakes, and

thence to the Mississippi and the mountains of the Carolinas. "From the Adirondacks they learned the art of husbandry" and became tillers of the soil; and when they were in the height of their glory they discovered some rays of light from the sun of civilization that peered through the dark shadows of paganism by which they were surrounded.

In the County of Onondago, when first settled by civilized man, numerous traces of a former occupation were discovered, which were then regarded as the works of the Indians; but they proved to be those of the French who, in the seventeenth century, were endeavoring to impress upon the pagan nations the blessings that came from the religion of Christ.

Notwithstanding the Iroquois were the most powerful and intellectual people north of the Montezumas, and that they displayed great energy, especially in war, and held much of the territory in eastern and central New York, I can learn of no mounds or military fortifications similar to those found in Western New York or that section lying between the southern shores of Lakes Erie and Ontario and the head waters of the Allegany River. No nation of Indians within the records of history has ever built structures that would stand the ravages of the time that has passed away since the foundations of many of the works located in Western New York have been laid. The great tree referred to by the Indian Chief, that stood partly upon the mound at Coldspring, and also the

giant pine whose topmost branches could overlook
the magnificent valley of the Cassadaga and the
great chain of hills that flank its eastern border,
present, in unmistakable language, the true history
of its age; a history, from infancy to old
age, unfogged with theories concerning " Behring's
Strait, or the lost tribes of Israel, " or from what
part of Europe the progenitors of these pure native
Americans originated, but an annual record is made;
and, in the case of one of these trees, there is figured out the vivifying influence of more than eight
hundred summers since it raised its head from the
midst of the intrenchment surrounding that ancient
field of gore.

If the Indian tribes that occupied the vast territory south and west of the great lakes thousands
of years ago, mined for copper at Lake Superior,
and for mica in the mountains of North Carolina,
and built great sepulchral mounds in which to deposit their dead, it would certainly seem that they
would have continued this practice to some extent
since they have become known to history. Since
the confederation the Iroquois are known as the
most intelligent of all the Indian nations, and they
are very remarkable for keeping sacred in their
remembrance the history and the great deeds of
their ancestors. It is admitted by all of the intelligent Iroquois with whom I have conversed, that
the people who constructed the mounds were an
unknown people, and that tradition interposes no
beams of light to extricate them from the confusion
which has been wrought by time. Dr. Wilson, an

educated Iroquois and a graduate of a Medical College, paid great attention to the origin and the antiquity of his people, and also to the builders of the tumuli upon the reservation where he was born; but he leaves the solution of the origin of those mounds enshrouded in obscurity. It is known to all investigators of history that the Indians are superstitious in regard to disturbing the dead; hence it has been argued that their desire to retain their mausolea undisturbed is an evidence that they regard them as containing the remains of their own people. They have learned that the mounds were made to cover the dead of an ancient people, and that their desecration would offend the Great Spirit.

It is the general opinion of archæologists that the Mound-Builders were worshippers of the sun, and that many of their most imposing structures had a religious significance, upon whose topmost limits they offered sacrifice to the god of day.

RELIGION OF THE INDIANS.

The Indians who occupied Western New York for a long period before the epoch of American history, had a theology that they have handed down to succeeding generations with religious veneration. The religious system of the Iroquois, notwithstanding it originated amongst untutored savages, has elements of great sublimity. It sees God not in a personality, but as a great and loving Spirit whose extended arms encircle the universe. In some respects it bears a resemblance to the religion of the ancient Greeks who peopled the in-

visible world with millions of departed spirits.
The Indian believes that the Great Spirit controls
the motions of the stars, forms the clouds and
places them in positions where they may send
down gentle showers to vivify the growing corn ;
that He watches their every interest with fatherly
care ; and that He will escort them to the charming
hunting grounds " beyond the river of death, "
where beautiful birds make resonant the hills and
the valleys with their enchanting songs. They
also believe that the Great Spirit has peopled the
invisible world with deer, bear, and all other animals useful for food, and has endowed the good
Indian with a capacity to climb the rugged hills
and never tire, and to sport upon the lakes and the
rivers that would never fail to supply them with
fish.

FESTIVALS OF THE IROQUOIS.

For the countless blessings which the Great
Spirit confers upon them, they meet at stated intervals to do Him honor.   Six annual festivals were
customarily held by the Iroquois.   The first was
held when the melted snow left bare portions of
the earth within the wilderness, and when the
rigors of winter had passed away.   This festival
was called the maple dance and was performed to
do reverence to the Great Spirit for the gift of the
maple, and for so tempering the atmosphere that
they could draw from its side its sweet waters.
This maple festival was confined to no particular
day, but the time was appointed when they assem-

bled for the confession of their sins and for return-
ing thanks to their Spiritual Divinity for the preservation of their lives during a cold and rigorous
winter. All of their festivals were preceded by a
meeting of the tribe in order that each one might
repent of his past sins and determine that during
the subsequent year he would avoid the snares and
pitfalls into which he had fallen in the past. At
the time of these meetings one of their leading
theologians would take in his hand a "string of
white wampum" and, whilst facing the audience,
would confess his various infirmities. The wampum was then passed from one to the other of those
assembled, until it reached every person in the
audience. The promise of regeneration was considered of no binding effect, unless sealed by the
wampum which was held in the hand. Those fes-
tivities were commenced with an opening address
made by one of their leading orators, whose eloquent words are treasured up in their memories.

Another festival held by the Iroquois was to
thank the Great Spirit for the return of spring, and
to bless Him for unfolding the buds upon the trees
and for decking the woods and the meadows with
variegated flowers, and also to invoke His blessing
upon the seed they had placed in the ground, that
it might there be raised to life and produce a
bountiful harvest. This vernal festival was regarded as of great importance, for they supposed
that the growth and the development of corn and
of all vegetable life was the gift of the Great
Spirit.

THE GREEN CORN FESTIVAL.

The " green corn dance " is considered of great importance with these aborigines of America, and is annually practised at the present day by the small remnant of Iroquois now occupying a narrow strip of land on either side of the Allegany River.

The green corn festival is very ancient and has been observed by these children of the American wilderness from a time that reaches into the eternity of the past, far beyond their remotest traditions. Since corn with them was regarded as "the bread of life, " it is not unreasonable to suppose that they would render thanks to the Great Spirit of the universe for tempering the seasons so as to produce in their time this sweet and delicious food.

The belief in a presiding Deity is indigenous to the nature of man. All nations have a god; but some, through ignorance of the laws governing the universe, have converted him into a demon of a malignant spirit. But the Indian idea of God is marked with sublimity. He is the tender and loving Father Who watches the interest of His children with the care bestowed upon the infant reposing in its mother's arms; and, whilst He continually holds in His hand the scales of eternal justice which He metes out to every son and daughter of the forest, He inflicts upon the wicked and upon those who have gone astray no sanguinary punishments, but holds out the hand of forgiveness to the worst of sinners, after they have been scourged with the whip of justice tempered with mercy.

The Indian believes that when the seasons have been unfavorable to the growth and maturity of their corn they have offended the Great Spirit; hence they increase the number of days devoted to their harvest festival which takes place when they can behold the ripened corn disclosing its golden fruit.

ANCIENT PRAYERS REPEATED AT THEIR FESTIVALS.

In the midst of one of the dances peculiar to the green corn festival, a prayer to the Great Spirit is repeated, which prayer, unaltered, has been handed down from a remote and unknown period in the past.

The assembled multitude, with heads bowed, listen with marked attention to the words of the orator, whilst he delivers the prayer rendered so sacred; for it was the production of their ancestors whose glorified spirits are now free from the shackles of the flesh and are hovering around the eternal throne of their loving Father. This prayer has been thus translated:

"Great Spirit in heaven listen to our words. We have assembled to perform a sacred duty, as thou hast commanded. This institution has descended to us from our fathers. We salute Thee with our thanks, that Thou hast caused our supporters to yield abundantly.

"Great Spirit our words continue to flow towards Thee. Preserve us from all danger. Preserve our aged men. Preserve our mothers. Preserve our

warriors.   Preserve our children.   Give wisdom
to the keepers of the faith, that they may direct
these ceremonies with propriety.   Strengthen our
warriors that they may celebrate with pleasure the
sacred dances of Thy appointment.

"Great Spirit: the council here assembled, the
aged men and women, the strong warriors, the
women and children unite their voice of thanks-
giving to Thee."

In some of the ceremonies of the Iroquois
superstition plays a conspicuous part.   During one
of their special prayers, always delivered at certain
of their festivals, they burn incense as an offering
to, and to attract the attention of the Great Spirit.
During the delivery of this prayer the orator scat-
ters some leaves of tobacco upon the embers, and
as the smoke arises the voice of the speaker is
drowned by the ejaculations of the audience.
This prayer, though pregnant with superstition,
has been translated into English and reads thus:

"Great Spirit who dwellest alone, listen now to
the words of Thy people now assembled.   The
smoke of our offering arises.   Give kind attention
to our words as they arise to Thee in the smoke.
We thank Thee for this return of the planting
season.   Give us a good season that our crops may
be plentiful.

"Continue to listen for the smoke still arises.
Give strength to us all that we may not fall.   Preserve our old men among us and protect the young.
Help us to celebrate with feeling the ceremonies
of this season.   Guide the minds of Thy people

that they may remember Thee in all their actions. "

The circumstance that the modern Indians in Western New York have never constructed great mausolea wherein to bury their dead, and that they have a religion bearing but a slight resemblance to that supposed to have been the religion of the Mound-Builders, and that not even a feeble ray of light has reached us from their remote traditions, would indicate that they are not of the same people; and that the Mound-Builders were far above the Indians in civilization and the arts, or more especially in ability and disposition to perform great labor and to encounter the rugged obstacles of nature.

WAS SUPERSTITION THE CAUSE OF THE GREAT NUMBER OF FORTIFICATIONS IN WESTERN NEW YORK?

That the Mound-Builders were a warlike and superstitious people is shown from the great number of works that have long stood the ravages of time. That superstition had much to do in constructing the many military fortifications in Western New York is apparent; and as it has always had much to do in all the wars carried on by civilized people, and as every pagan nation is pregnant with superstition, make it apparent that the Mound-Builders were subject to its terrible influence.

Commencing at the southern shores of the northern lakes and extending southward a hundred miles or more we find a greater number of military works than in any other section of the United States, in proportion to the amount of territory.

In parts of the state of Ohio and the Mississippi valley are many that are built on a more magnificent scale, but in point of numbers the lake regions far exceed them. It is my opinion that the terrible conflicts which produced such a great number of fortifications as we find in Western New York, were instigated by superstition which, with all nations and peoples, is the fountain head of war and cruelty. It seems impossible that a people conditioned as they must have been would have spent so much labor in building such monuments and fortifications that have resisted the action of the elements for perhaps thousands of winters, unless they were fired by a veneration for some terrible superstition, or for the dread ministers by whom it was interpreted. The religions of the past either directly or indirectly were connected with war and cruelty; and no religion of which we have any knowledge is exempt from the terrible charge. Mahomet the father of a religion whose followers outnumber the hosts of Christendom was by nature endowed with great benevolence, but in the name of his god he could see the quivering flesh of men, women and children torn and lacerated, and could then sing praises to Allah for the destruction of unbelievers. The Christians, in the wars of the Crusades, in order to get possession of the Saviour's sepulchre whitened the soil of Europe with ten millions of human skeletons and committed cruelties which language is inadequate to name. These bloody wars were instigated to get possession of the tomb where Jesus was buried. After these terrible wars had been in progress for years the Christian army battered the walls of Jerusalem and entered the holy City in possession of the Saracens. A celebrated historian ("Gibbon's Decline and Fall of the Roman Empire") in describing the scene says: "A bloody sacrifice was offered to the God of the Christians resistence might provoke, but neither age nor sex could mollify their implacable rage. They indulged themselves three days in a promiscuous massacre and the infection of the dead bodies produced an epidemical disease. After seventy thousand Moslems had been put to the sword, and the harmless Jews had been burnt in their Synagogues, they could still retain a multitude of captives which interest or lassitude persuaded them to spare. Of these savage heroes of the cross Tancred alone betrayed some sentiments of compassion. Yet we may praise the more selfish lenity of Raymond who granted a capitulation and safe conduct to the garrison of the citadel. The holy sepulchre was now free and the bloody victors prepared to accomplish their vow. Bareheaded and barefooted with contrite hearts and in an humble posture they ascended the hill of Calvary amidst the loud anthems of the clergy, kissed the stone which had covered the Saviour of the world and bedewed with tears of joy the monument of their redemption."

A religion may be pure and holy and possess every element necessary to lead mankind to the sublimity of virtue, but when its subjects become entangled in the web of superstition its glorious

precepts are trampled in the dust and its victims
become lost to every principle of humanity. This
terrible demon superstition no doubt took possession of that numerous people that once held empire over the American Continent, and led them to
the verge of insanity. Mica, as previously referred
to, was no doubt in some mysterious way connected
with their superstitions, for nearly every mound
which has been opened has disclosed blocks of that
substance. Other relics which were regarded by
those people as containing great value were buried
with the dead. Every discovery made carries con-
viction that they believed in an immortal life beyond the grave, and also in the resurrection of the
body, else why should they entomb in every mausoleum for their dead their sacred relics and implements of warfare? They believed they would
rise again in a bodily form and use them as heretofore.

Every pagan nation has some idea of an immortal life, and all the great structures found in Europe
or America show evidence that that belief was
prominent in all their works. Had it not been for
" their longing after immortality " the foundations
of the mighty Pyramids would never have been
laid. One of the ancients speaking eloquently on
the subject of immortality, says:

> " The stars shall fade away,
> The sun himself grow dim with age,
> And nature sink in years ; .
> But thou shalt flourish in immortal life,
> Unhurt amidst the war of elements,
> The wreck of matter and the crush of worlds. "

From the appearance of the mounds and their location and mode of construction found in Western New York, I am led to believe that at the time those cruel and devastating wars were waged the assaulting party came from the south; and it is not improbable that there might have been a northern and a southern confederacy, and that policies might have been in some way connected with it. As I have suggested in a previous chapter, much of the copper mined at Lake Superior was brought down the lakes and transported to the south by way of the Conewango, the Cassadaga, and the Chautauqua Lake. There might have been a contention with regard to lake navigation and perhaps to northern production.

From the great number of ancient forts found in the north I am led to believe that the northern army was driven towards the lakes and made a stand in the places where the ancient forts are now found. That they held their ground for a long time is shown by their numerous fortifications. I have no doubt that traces of scores of these works have been discovered in the counties of Chautauqua and Cattaraugus alone. It must have been impossible that such a great number of extensive works could have been constructed short of many years, for we have no definite knowledge that they understood the use and manufacture of iron. To have removed such great piles of earth without the use of iron shovels, it seems would require the labor of an army for a lifetime. Embracing several counties in Western New York, I believe more yards of earth have been removed for military purposes than were removed by the vast armies connected with our late civil war. If the grounds whereon were fought the battles of Fredericksburg, Spotsylvania and Petersburg were exposed to the action of time that has passed away since those ancient works were constructed, I have no doubt but that many of these bulwarks would have become obliterated.

It is painful to write of a people that once occupied the vast territory comprised in sections where the footprints of ancient man is so well defined, and to know nothing of their history or their origin, or whence they came, or whither they went; and we are puzzled to conceive how such armies as must have been in the field for many years at least, could have been sustained. They must necessarily have interfered with the cultivation of the soil to a great extent. Some suppose that they lived on game, but that must have been soon exhausted. If my theory that they tamed the mastodon and used him as a beast of burden is correct, they might have transported their food from other sections by the aid of that powerful beast far away from the tide of war; but the manner in which they lived and who was the victorious party, are facts shrouded in the darkness of the long ago.

"That would be a great book which could detect and eliminate the facts connected with the latent movements and actions of man since he first walked upon the surface of this sublunary world. But it is a fact unquestioned by philosophers that our acts and words produce effects that must reverberate through all time so the whole of futurity would be different had our words never been spoken or our deeds never been enacted. The pulsations of the air once set in motion by the human voice cease not to exist with the action that gave them rise. Strong as they may be, in the vicinity of utterance their quickly attenuated force soon becomes inaudible to the human ear. But the waves of air thus raised perambulate the earth and oceans' surface, and in process of time every atom of its atmosphere takes up the altered movement due to that infinitesimal position of primitive motion which has been conveyed to it through countless channels, and must of necessity influence its path throughout its future existence. Every atom impressed with good and evil retains at once the motions that sages have imparted to it, mixed and combined in ten thousand ways with all that is worthless and base. The air is one vast library on whose pages is forever written all that man has ever done or even whispered. There in their unerring characters mingled with the sighs of mortality stand forever recorded every act ever performed and every deed ever committed. " This great book indelibly printed on air and earth and ocean is the only one that contains a reliable history of that long lost race of men.

CHAPTER VI.

THE ALLEGANY RIVER.—ANCIENT WORKS ON ITS
BANKS.—THE MODERN INDIANS—THEIR WARS.

The Allegany River presents to the lover of
nature more interesting characteristics than any
other river in the world.   One of its most remarkable features is the purity of its waters fed, as they
are, by mountain streams for three hundred miles;
hence it received its name, Allegany, from the
Seneca Indians, meaning "Fair Water."   It flows
to the south, and its current and direction are so
winding that there is no point of compass to which
it does not direct its course.   The banks are exceedingly wild and rugged, and its precipitous
bluffs and hills frequently rise to the height of
three hundred feet.   This majestic river will always
have its charms, although associated with all the
horrors of savage warfare.   Its clear crystal waters
gliding swiftly over its shining pebbles make it
one of the most lovely streams that ever glistened
in the light of the sun.

The Allegany River has always been venerated
by the modern Indians.   A life on its banks and
the light canoe is the red man's paradise.   Black-

snake made his home near its waters and selected the spot to lay his bones. Cornplanter, the scarred veteran of a hundred battles, with Blacksnake his successor slumbers now in deep repose on the banks of his own beloved Allegany.

At one of my interviews with Governor Black-snake, in the year 1853, he said there formerly existed a number of tumuli in the vicinity of Corn-planter's Island, but from ancient cultivation they were now leveled down. Cornplanter, the owner of the Island that bears his name, was in early life a skilled and cruel warrior. He allied himself with the French about 1750, and in 1755 fought as a master spirit the battle of Braddock's Field, a battle which proved a terrible disaster to the English forces, there being 700 soldiers killed and wounded together with 40 officers. General Washington at that time was an officer in Braddock's army, and remembered to the day of his death the fiery spirit of that antagonist. They were both young men at that time; the one treating his prisoners with fiendish and savage cruelty, and thrusting into the flesh of his naked victims splinters of wood saturated with some combustible substance which, when set on fire, their screams and agonies were drowned by the savage yells of the victors; the other treating his prisoners with humanity, binding up their bleeding wounds and watching over them with tender care.

Notwithstanding the Iroquois were numerous at the time the country was first explored by the Europeans, I think the ancient population was much greater, for they have been traced from its head waters composed of mountain springs to its junction with the Monongahela. Most of the mounds that formerly existed on its banks have been leveled down and are now known only in history. Persons who have traveled down this winding river will recollect the Oil Creek that pours its turbid waters into the current and forms one of its prominent tributaries. As we ascend this oleaginous stream to the distance of about four miles, we find singular traces of ancient works very different from any that have been referred to heretofore. On what is known as the Buchanan farm, is an ancient shaft which was formerly sunk to the depth of about thirty feet. This was done for the purpose of collecting oil. As the oil is less dense it readily stands upon the surface of water and was thus obtained. A half mile up the stream occur numerous pits which were formerly dug for the same purpose. As the French traversed nearly all the streams bearing in a southern direction on their way from the great lakes to Fort Duguesne, (now Pittsburgh), it was generally believed that they, in connection with the Indians, dug those pits for the purpose of supplying themselves with oil. The fact that iron implements of French origin have been found in various places on Oil Creek, makes it very reasonable to attribute them to that people; and we have no doubt they gathered oil there to some extent. But that those pits were dug by the French is out of the question, when we take into consideration that forest trees have been found growing upon them, that had waved their branches amidst a hundred winters before Columbus pointed the prow of his little ship into the unknown and trackless ocean towards the shores of the new world. In the year 1861 I saw tools found in different places on the creek, which were composed of native copper, one of which weighed several pounds. It was something like a drill, rather flat pointed at one end and appeared to have been hardened. The circumstance of finding copper in that section is evidence, almost beyond a doubt, that those pits, or many of them at least, were the work of the Mound-Builders as they are the only ancient people that made use of copper to any extent. They used it for various mechanical purposes, as well as to bury it with their dead.

As we descend the Allegany from the mouth of Oil Creek to the village of Franklin the noble river increases in beauty and grandeur. On the left the banks are steep and rugged, and its hills covered with evergreen trees and shrubs tower towards the heavens to the height of several hundred feet.

The village of Franklin is located at the mouth of French Creek whose waters form one of the greatest tributaries of the Allegany River, where have been found many interesting works made by those ancient inhabitants; but from the long time this country has been settled and cultivated by the whites nearly all the mounds and fortifications are lost to the archæologist. "On the summit of a high hill overlooking the river and stream," says Mr.

Babbit, " is an Indian grave built with great care, and much labor has been expended in its construction. " The grave was no. doubt that of some aged chief who had spent his days in a bloody struggle for universal empire, and who preferred to lay his bones amid the rocks and solitude of the mountain. His wild and poetic religion no doubt taught him that the sleepless spirit of the warrior would hover around him, and that he as a sentinel could point to his tribe the true path that would lead to safety and glory.

Descending the river to the neighborhood of Foster's Island we find one of the most interesting specimens of ancient workmanship ever found in this country. A large rock stands on the left margin of the river, known to all travelers and pilots as the Indian God. This rock undoubtedly to some extent records the history of a departed and forgotten nation that once flourished upon its banks and sported upon its waters. Many figures are carved upon it. Among them can be distinguished a snake, a turtle, an eye, an arrow and a sun. Several others more prominent resemble different parts of the human frame. The inscriptions on this rock have been regarded by Archæologists as the work of the primitive inhabitants. These symbols or hieroglyphics were no doubt made to chronicle some important event connected with their wars or superstitions; a stone-graven story of man's forgetfulness.

As we glide down the Allegany River the scenery that presents itself becomes more beautiful and interesting at every step. Majestic hills covered with huge rocks mock the sunlight, and its elegant towns and cities add new charms to the scene. The villages located on either side are eloquent with the sound of the hammer, the buzz of the saw, the prosperity and activity of its bankers and merchants, together with the voice of the husbandman as he distributes the golden harvest. School-houses are seen in great numbers, preparing the rising generation to become useful and intelligent members of this great republic and make it foremost among the nations of the earth. The voice of the locomotive is heard in every direction and reverberates from hill and mountain. Magnificent cars loaded with living freight are carried through valleys and over mountain streams at the rate of forty miles an hour. Steamboats majestic in appearance ply upon the bosom of the river, defying its powerful current and breathing fire and smoke. Go back thousands of years; yea, before the foundations of Rome were laid, and the banks of this river were alive with human beings engaged in the business of active life and building sepulchral monuments to cover the remains of their venerated dead. The mounds and fortifications that were once very numerous on the banks of the Allegany are now from a long cultivation nearly obliterated. Go back a hundred years or more and this river was occupied by the Seneca Indians, the most powerful and warlike of the Iroquois confederacy. Cornplanter had allied himself with the French and fought many a bloody battle between the old fort at Franklin and the mouth of the Monongahela River. One of the most prominent and daring characters connected with those Indian wars was Captain Samuel Brady who was reared amid the excitements of a section exposed to the cruelties of savage warfare. In early life his father and brother were murdered by the Indians, and they suffered the most cruel tortures. From that moment Brady took a solemn oath of vengeance against the Indians for the murder of a brother and a beloved parent. In 1779, near the mouth of Redbank Creek, Captain Brady met a large force of Indians led by Cornplanter. The battle soon terminated in favor of the whites. Many of the enemy were killed or wounded; others fled to the river that was in full flood and dashed into the powerful current; few succeeded in reaching the opposite shore, for one by one their strength gave way and they sank into the wave of death. Cornplanter almost alone braved the stormy river and saved his life. From that time the brave, cruel and revengeful spirit of that daring and savage chieftain began to falter. From 1791 to the day of his death he devoted his life to peace and used all his influence to restrain the vengeance of his people. He entertained the highest respect for Gen. Washington and always called him " the great counselor of the thirteen fires. " He visited him frequently, during his Presidency, on business connected with his tribe.

A writer that visited Cornplanter, in 1834, and had an interview with him says: " I saw this ven⁖

erable and aged chief about a year and a half
before his death. I thought of many things while
seated near him beneath the wide-spreading shade
of an old sycamore, on the banks of the Allegany;
many things to ask him: the scenes of the Revolution, the generals that fought its battles and conquered the Indians, his tribe, the Six Nations and
himself. He was constitutionally sedate, was never
observed to smile, much less to indulge in the lux-
ury of a laugh. When I saw him he estimated
his age to be over 100 years — 103 was about his
reckoning of it. This would make him nearly 105
at the time of his death. His person was much
stooped and his stature was far short of what it
once had been. Time and hardship had made
dreadful impressions upon that ancient form. His
chest was sunken and his limbs had lost their size
and become crooked. His feet too had become
deformed and haggard by injury. Most of the
fingers on one hand were useless; the sinews had
been severed by a blow of the tomahawk or scalp-
ing knife. How I longed to ask him what scenes
of blood and strife had thus stamped the enduring
evidence of its existence upon his person. But to
have done so would in all probability have put an
end to all further conversation on any subject, so
I had to forego my curiosity. He had but one eye
and even the socket of the lost organ was hid by
the overhanging brow resting upon the high cheek
bone. The remaining eye was of the brightest
and blackest hue. Never have I seen one in young
or old that equalled its brilliancy. Perhaps it bor-

rowed lustre from the eternal darkness that rested on its neighboring orb. His ears had been dressed in the Indian mode — all but the outside ring had been cut away. On one ear this ring had been torn asunder near the top and hung down like a useless rag. He had a full head of hair white as snow, which covered a head of admirable shape. " Cornplanter died on the 7th of March, 1836, at the age of about 105 years, and for a long time previous had been a man of peace. Notwithstanding his deeds of cruelty in early life, growing out of his superstitions, (for he was always superstitious), his latter days were devoted to humanity and truth. He died believing in the religion of his fathers and looked with an eye of faith towards that beautiful hunting ground believed in by his people.

> " But thinks admitted to that equal sky,
> His faithful dog will bear him company. "

When this celebrated Indian died there fell the last great chief but one who will ever preside over that once powerful confederacy, the Iroquois. Their cycle now is nearly filled and shall soon have passed away forever. On the banks of the majestic Allegany that has had such charms for so many nations and peoples, now repose in silent dust the bones of the Mound-Builder, the Indian and the white man. If the doctrine taught by a large portion of the Christian world will prove true, what a wonderful scene will be presented when the "angel shall sound the trumpet. " In idea one of Scotland's most sublime poets has beautifully described it :

> "Now starting up among the living changed,
> Appeared innumerous the risen dead.
> Each particle of dust was claimed ; the turf
> For ages trod beneath the careless foot
> Of men, rose organized in human form ;
> The monumental stones were rolled away,
> The doors of death were opened ; and in the dark
> And loathsome vault, and silent charnel house,
> Moving were heard the mouldering bones that sought
> Their proper place. Instinctive every soul
> Flew to its clayey part ; from grass grown mould
> The nameless spirit took its ashes up
> Re-animate ; and merging from beneath
> The flattened marble undistinguished rose
> The great, nor heeded once the lavish rhyme
> And costly pomp of sculptured garnish vain.
> The Memphian mummy, that from age to age
> Descending, bought and sold a thousand times
> In hall of curious antiquary stowed,
> Wrapped in mysterious weeds the wondrous theme
> Of many an erring tale, shook off its rags ;
> And the brown son of Egypt stood beside
> The European, his last purchaser.
> In vale remote the hermit rose, surprised
> At crowds that rose around him, when he thought
> His slumbers had been single ; and the bard
> Who fondly covenanted with his friend
> To lay his bones beneath the sighing bough
> Of some old lonely tree, rising was pressed
> By multitudes that claimed their proper dust
> From the same spot ; and he that richly hears'd
> With gloomy garniture of purchased woe,
> Embalmed in princely sepulchre was laid
> Apart from vulgar men, built nicely round
> And round by the proud heir who blushed to think
> His father's lordly clay should ever mix
> With peasants' dust — saw by his side awake
> The clown that long had slumber'd in his arms. "

As I have occupied considerable space in describing the Allegany River and the Indians that formerly resided on its borders, it may appear to some of my readers as an uncalled for digression,

but as those ancient people made the river a highway of travel from the great lakes to the south,
and cultivated its interval lands and laid their
bones upon its shores, I shall frequently have occasion to give it reference.

On the Allegany, about eleven miles above the
city of Pittsburg, on ground now occupied by the
beautiful village of Mechanicsburg, were formerly
several tumuli which contained a large amount of
human bones. A mountain stream of pure water
enters the river about one hundred rods above the
village. Near the bank of this creek was formerly
a large military fortification that is now nearly
obliterated. Arrow heads, lances, and stone axes
have been found in large quantities. That a battle was fought there by the ancient warriors is beyond a doubt. As several battles were fought in
that neighborhood by Capt. Brady and the modern
Indians, it is probable that some of the relics found
were of Indian origin. In the spring of 1875 I
had an interview with an aged man born and reared
in that vicinity, who claimed to have found numerous copper implements of curious workmanship on
the opposite side of the river, on a small plateau
of land cultivated by him for many years; but,
being ignorant of their value, he had given them
away and they were lost.

As we pass down the Ohio River from the city
of Pittsburg to Wheeling, West Virginia, numerous mounds and fortifications have been discovered,
but for the want of reliable information I am unable
to describe them.

At Martin's Ferry, nearly opposite the city of Wheeling, is a more extensive ancient work than I have yet had occasion to mention. It is located on the second terrace which is about fifty feet above high water mark. This work is nearly square with the two corners facing a hill to the westward, rounded off. It is elevated some ten or twelve feet above the surrounding plateau and occupies about three acres. It is, in fact, a pile of earth artificially raised with gentle slopes on either side. This embankment on the top is as level as the floor of a house, and to all appearances has changed but little since its first construction long ages ago. The soil of which it is composed is very fertile and has been held no doubt by a network of roots of trees and shrubs that have been growing upon its surface for thousands of years. Near one of its corners, so as to be almost connected, is a mound much more extensive than any that has been discovered in the State of New York. In 1875, while spending several days in that locality, I made an estimate of about four hundred feet for the circumference of the mound, and for height about thirty feet. Large trees are growing around and upon the banks of this tumulus, which seem uncommonly thrifty. The mound has been partially opened many years ago, but from present appearances the centre has never been disclosed. I was informed by several parties residing there that human bones, copper implements and other relics were found.   The Rev. J. P. Maclean, in a series of articles written for the " STAR OF THE WEST" is of the opinion that works of that char- acter were devoted to some kind of amusement, and says " it is doubtful if a people can become so subdued as to be restrained from all amusement. Amusement is a necessity ; it is an element of man's nature.   If there are those that are strangers to such a sense they are to be pitied and their con- dition shunned. "   With regard to amusement, no person who has reflected upon the .nature of man will disagree with the Reverend Gentleman.

If this work were devoted to some peculiar amusement it would seem singular that they should build such stately tombs by its side.   Perhaps it was made to cover the remains of some prominent actor who was uncommonly skilled in performing some interesting drama of human life, or of some Shakspeare who had won an immortal renown in shaping some comedy or tragedy and given it a reputation and a name.   One thing is true, that this singular monument is located amidst the most beautiful and enchanting scenery that we ever beheld.   The last time I saw it was in the month of May, just as the sun was going down.   The beautiful Ohio moved rapidly along its course, re- flecting back the rays of the god they worshipped, and on the opposite side of the river a precipitous hill stood majestically defying the rushing waters. About a mile below could be seen the city of Wheeling and steamboats pouring out great vol- umes of smoke.   Directly opposite the city one of the largest and most splendid islands in the Ohio presented itself to view, dotted with streets and buildings. Whilst standing over the remains of the long buried dead I longed for the eloquence of a Byron to depict the enchanting scene :

> " Clime of the unforgotten brave !
> Whose land from plain to mountain cave
> Was freedom's home or glory's grave,
> Shrine of the mighty ; can it be,
> That this is all remains of thee. "

Whether this great earthwork was built for amusement or not, is very uncertain; but from its mode of construction and its general characteristics, it is highly probable it was devoted to a far different purpose, a purpose in some way connected with religion. In every section of the country where the ancient works are found many of them show a religious significance. The religious element is so strong in the nature of man that it does and always has shown itself, not only with the most degraded, but with the most cultivated and intellectual people everywhere. If every religion now extant amongst the different nations of the earth should forthwith be abolished, in less than four weeks the world would be in arms for the construction of new ones. Mankind will always have some invisible deity that to them will be an object of worship. The inhabitants of Mexico and Central America worshipped the sun, falling upon their knees to do him reverence when he arose majestically scattering his golden rays upon the mountain tops and fertile hills. It is highly probable that those ancient inhabitants built their earthworks for the purpose of doing reverence to some imaginary god, or to those brilliant luminaries that glisten in heaven's archway. It is not impossible that this great earthwork was devoted to human sacrifice. If such was the case, we can imagine terrible scenes to have been enacted there,—scenes that would chill the blood. Men, women and children were offered upon the altar and their hearts torn from their bleeding bodies by the priests of the bloody faith; all done in honor of or to appease the wrath of some avenging deity.

Up the river Ohio, for the distance of several miles above Martin's Ferry, is a broad flat of alluvial land which, to all appearances, was once extensively cultivated — done very likely by the people who built those mounds and fortifications so extensive on almost every interval of land on the beautiful river. The flat above referred to once contained a large number of tumuli from fifty to seventy-five feet in diameter, and also large quantities of works of ancient art, such as are usually found in those great repositories of the dead. A single mound now exists on that plateau of land, which has never been desecrated by the white man. One individual of eccentric characteristics willed to lay his bones on this mound in company with those ancient dead, and to-day a splendid marble slab marks his narrow bed above the mouldering skeletons.

As we pass down the Ohio River from the city of Wheeling the scenery is grand and interesting. The valley presents a wide and beautiful expanse.

Islands are frequently met covered with splendid foliage. The valleys are rich with verdure on either side, and in the dim distance hills rise in shadowy outline, as if uncertain where to limit their boundaries. Crystal streams meander down the valleys and mingle their waters with those of the Ohio.

THE GREAT SEPULCHRAL MOUND AT GRAVE CREEK.

Twelve miles below the city of Wheeling, on the West Virginia side of the river, occurs a large interval of land of wide expanse, which is flanked to the eastward by precipitous hills located a mile or more from the Ohio river. The village of Moundsville is located here. It was formerly known as Grave Creek, on account of the sepulchral mounds which it once contained. As the river traveler passes down the Ohio until he arrives at Moundsville, he is there presented with one of the most enchanting locations of Nature's handiwork. His attention is at once directed to a singular hill, symmetrical in form, which towers to the height of seventy feet. The hill is located amidst a level plateau that contains more than a thousand acres. This imposing monument is artificial and is the great sepulchral tumulus of Grave Creek, one of the largest of the kind ever constructed by human hands. About half a mile away is a mountain stream that flows rapidly down, bounding over the rocky cliffs till it meets the Ohio, where they mingle their waters like a stream of life flowing on to the gulf of eternity.

This mammoth mound is located about half a mile from the Ohio river and some forty feet above its shores. Its base is a true circle with a diameter of more than twenty rods and a circumference of over sixty. It is conical in form, and before the top was leveled off it was about eighty feet in height. The time that has passed away since this great work was built is shrouded in mystery. Some suppose it has stood for thousands of years, others regard it as more recent; but it is not improbable that since its construction Rome, Greece and Thebes have fallen, and that their gorgeous palaces, rotundas, obelisks and colosseums have gone down in a common sepulchre. Although this great tumulus can claim no name in history, it stood defying the frosts of winter, the roar of the hurricane and the corroding elements of time when great nations and peoples in their palmiest days of prosperity occupied magnificent temples of granite and marble, which are now a heap of ruins.

In the year 1838 this tumulus, after having attracted much attention from those curious to know its contents, was excavated by a Mr. Tomlinson who claimed a legal title over the Mound-Builders by adverse possession. He not only held the land but also the great mound that cost a million hard days' labor to construct. The course pursued in penetrating the mound was to commence at the base in the form of an arc. The soil being composed in part of clay and being perfectly dry showed no signs of caving in, so they readily reached the centre and, to the astonishment of all concerned, came to a room or vault twenty feet square, the bottom of which was covered with a sort of dark material made from decayed timber that had long since been reduced to ashes by the destructive influence of time. After removing some of the debris and making a search, they discovered two human skeletons in almost a perfect state of preservation. One was that of a male and the other that of a female. The two were separated the distance of about ten or twelve inches; they lay side by side in a horizontal position with their heads to the south. The arms of the corpses, when deposited in that dark and silent vault, were laid carefully beside the bodies, and they had ceased to move for centuries before Columbus dreamt of a Western Continent. Four bracelets made of copper artistically designed surrounded the wrists of the corpses. The rings or bracelets have the appearance of having been made of wire mechanically manufactured and bent in a circle in the same manner pursued by jewelers at the present time. On the inner side of each wire could be seen a perfect impression of the cuticle or exterior coat of the skin, where it united by some chemical condition growing out of the decomposition of the part with which it was in contact.

When I visited this mound in the spring of 1841 it was open for exhibition. Mr. Tomlinson, a gentleman of education and very much of an antiquarian and the owner and proprietor, spent considerable time in imparting the information desired, so far as was in his power. He had been to

a large expense in opening and preparing it for the
reception of visitors. He had caused the bones of
the two skeletons to be wired and put in a position
the most favorable for exhibition. These two
skeletons were the best preserved of any that have
ever been taken from any mound in this country,
and were calculated to open a new field of investigation. By the aid of them we can arrive more
closely at the size of that ancient people, and as
they were undoubtedly prominent personages we
can form a very reasonable opinion with regard to
their mental ability. The two were supposed to
be about the same age, perhaps about fifty years.
The male, judging from the skeleton, was a man
about six feet in height, or perhaps a trifle more.
The female was estimated to be about five feet
seven or eight inches in height with bones which
would denote a powerful frame. The teeth,
according to my recollection, were generally sound
and the alveolar processes had been but little
absorbed, indicating that they died in the hey day
of life. The intellectual and moral regions were
exceedingly well developed. Benevolence and
reverence were large, which would indicate a person possessing a strong religious nature. The organ
of philoprogenitiveness, or juvenile love, was uncommonly large; combativeness and destructiveness were only moderately developed. Such a
head would indicate a person of strong moral
feelings; one that would love the right and despise
acts of cruelty and oppression; would love the
young with more than ordinary mother's love, and

at the sight of a suffering child would be melted to compassion. If she had children she would devote her life to them with the bright anticipations of a parent's heart and the burning hopes of a parent's soul. Should sickness and death overtake them, tears that would bedew her inanimate countenance would never wash out the dismal dye of the gloomy pall, or erase from the tablet of her memory the loved one passed away. As strong religious faculties were developed in the head of this queen of the Mound-Builders, we would naturally suppose she would reverence some higher power than that of man; no doubt some invisible pagan divinity, a subordinate to the sun, to whom she would pray with the frenzy of a superstitious devotion. It is very likely if influenced by priests she might sanction the most revolting cruelties and trample those high and glorious qualities so bountifully bestowed by nature's God low in the dust. The head of the king (if he was a king) was less developed in the moral organs than that of the female. The animal organs were rather large and the intellectual fairly developed. Such an organization would give considerable force of character and, if devoted to politics, would be no more likely to jump the track of honesty and virtue than many of our members of congress, bank directors and members of the United States Senate, who live in a land and an age of civilization unknown to the builders of those wonderful tombs.

When I examined this great mound, in 1841, I had at that time paid considerable attention to the science of Phrenology and had frequently delivered lectures on that and kindred subjects, so it seemed appropriate to give those heads a thorough examination. The heads of those persons bore but a slight resemblance to the heads of the Indians; neither were they like the Celtic or the Saxon people; but they would fall but little below them in native ability. Whether phrenology as a science is always reliable, or not, one thing is true, that it is able to detect the intellectual and benevolent from the low and degraded savage, though the bones of both have been preserved in the ground for thousands of years.

It is said that man is to some extent the creature of circumstances; and the circumstance of visiting that magnificent tomb, in the year 1841, is very likely the cause of the existence of this book on " Ancient Man in America. "

The sight of those skeletons, so well preserved, which had slumbered in the arms of death for perhaps many thousand years, produced an impression that shall never be blotted from memory. I saw before me the bones of the dead whose true history can never be written — the leaders, no doubt, of a once numerous and powerful people who at one time roamed o'er the hills and valleys of a large portion of the American Continent — and longed to know their history; to talk about the great animals, now extinct, which lived when they lived; to know the history of their different forms of government; also to learn something of their religious institutions; whether their priests led them to deeds of blood and cruelty, as has always been the case with other Pagan religions when unenlightened; to know the object of building such great tumuli at so great an expense, whether it was purely on religious principles or in part political; to know if they worshipped idols, and, if so, if it had any influence in their downfall as a people; also to know if they sacrificed human beings on altars dedicated to their gods, or made propitiatory offerings to their invisible divinities and crimsoned their altars with human gore; or did they have pastimes like those of the olympian plain, or struggles in the gladiatorial arena, or epicurean feasts, or bacchanalian revels.

> "Still silent incommunicative elf !
> Art sworn to secresy ? Then keep thy vows;
> But prithee, tell us something of thyself—
> Reveal the secrets of thy prison-house;
> Since in the world of spirits thou hast slumbered,
> What hast thou seen, what strange adventures numbered?
> If the tomb's secrets may not be confessed,
> The nature of thy private life unfold;
> A heart has throbbed beneath that leathern breast,
> And tears adown that dusky cheek have rolled.
> Have children climbed those knees and kissed that face?
> What was thy name and station, age and race? "

The soil of which this mound was constructed had not been obtained in the immediate vicinity, but had been brought from the distance of half a mile or more. Large pits can now be seen from which it was taken. The soil is a kind of clay and loam peculiar to this locality and is well adapted to the purpose for which it was used.

This mound contains about three million cubic feet of earth. The question is frequently asked, how were such great earthworks constructed? Some suppose they used beasts of burden and hauled the material on some rude sled or cart; others that the work was all done by human hands and carried on the backs of the laborers to the place of deposit. If the latter view is correct, what an immense amount of hard days' labor was required to perform such a stupendous work. A cubic foot of earth will weigh nearly one hundred pounds. Suppose a laborer to carry one-half a cubic foot to a load, dig it with some wooden tool made in some rude manner, carry it to the place of deposit, and dump it on the mound on an average of about thirty feet from the base; when his day's work was ended, he would have traveled the distance of eight miles, one-half of which bending under a heavy load. Allowing him three hundred working days to the year, he would, at the end of the year, have carried but twelve hundred cubic feet; and if he should continue on without interruption for the term of two thousand years, he would have deposited about the three million cubic feet of earth of which the mound is composed. If the life of a single man should prove too short for such a labor, one hundred men could perform the work in about twenty-five years.

This great mound was not alone in this repository of the dead. A number of tumuli formerly existed there, varying in size from ten to thirty feet in height, and when excavated they all contained osteological remains which, on exposure to the air, quickly crumbled to dust. This locality is remarkable for the vast amount of relics which were found among the skeletons. Implements of copper and stone were very numerous, and mica

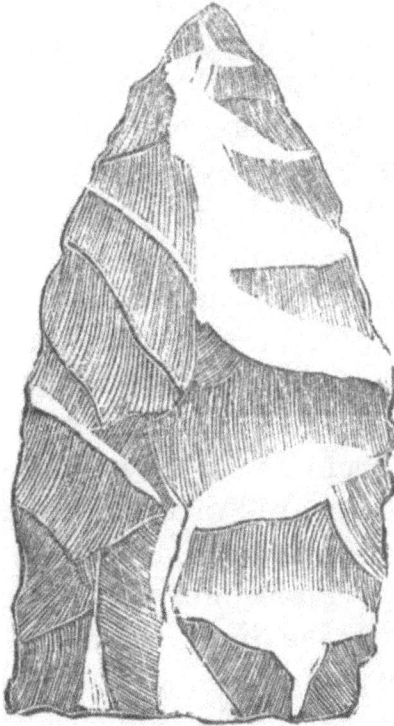

Fig. 12.—Relic composed of hornstone found in large numbers.

was found in larger quantities than in any other locality. In one of the mounds was disclosed an image wrought in sandstone, and for "artistic design" was in many respects superior to the one found on the Allegany River, near the mouth of Coldspring Creek. In the spring of 1841, in the month of May, in company with Mr. Webster we visited this ancient cemetery for the first time; and as we walked among these ruins we were inspired, as never before, with a thousand feelings and a thousand thoughts. We were standing on the dust of a long lost people, the last vestiges of a fallen empire already swept beneath the tide of time. Where were those people, and why should such immense labor be performed merely to cover the remains of a few, when the evidence was before us that it was the home of toiling thousands? Were the masses free, or were they in a condition of servitude? We cast our eyes upon the giant tumulus and the great forest trees growing upon its side gently bowed to the passing breeze and no answer came from them, and the crumbling bones were silent. The hills which had echoed back great words of orators, of kings and of priests, and the noisy storm of war, gave no answer. The sun that saw the honored dead entombed and the toiling multitude of men from year to year bowed down with heavy loads, heard their words and cheered them with the light of day, he too was silent and no answer came.

The question whether those ancient people were very numerous has long ago ceased to be a debatable subject. The State of Ohio alone has already disclosed more than twelve thousand ancient works, some of which are of giant size, and many hundred more, no doubt, formerly existed which were obliterated long before the foot of the white man touched her soil. This fact alone seems almost conclusive, that they cultivated the ground to a large extent and also in a methodical manner. The garden beds of the north-west, which were supposed to have been made by them, amply testify to the method pursued. These beds occur in the northern part of Indiana, extensively in Missouri and along the fertile banks of most of the rivers in the State of Michigan. The location selected for these beds was in the richest soil. Says Mr. McLean, "The lines of the plats are rectangular, as a general rule, but some are parallel and some semi-circular and variously curved, forming avenues differently grouped and disposed. They cover from ten to one hundred acres and frequently embrace even three hundred acres." The extensive works at Grave Creek, which must have employed thousands of men in their construction, attest almost beyond a doubt that they could have never been sustained by hunting and fishing, as was formerly supposed. The valley of this creek is composed of very fertile land and also the large flats on the borders of the Ohio, which extend for several miles on either side. It is very apparent that these lands were cultivated for thousands of years before the white man trod the Western Continent. Moundsville is now a thriving village located amidst the mouldering bones of a people around whose history will forever hang the dark veil of obscurity.

HOW THE VAULT WAS CONSTRUCTED—REMARKS.

The vault located in the centre of this mound was constructed with wood and then surrounded with earth. When first reached by Mr. Tomlinson the wood was found to be completely decayed; naught remained but dust mingled with the skeletons. It was reported at that time that the logs enclosing the vault had left an impression upon the surrounding soil so plain that the character of the bark could be readily distinguished. The top of this wooden structure was arched so that when the timber decayed the superincumbent earth would remain in its place. I was told, whilst visiting this mound in 1873, that from an examination of the impression made upon the soil by the bark (which was all decayed) some of the timber must have belonged to a species now extinct, and that no trees now existing present the same appearance. So much has been said and written with regard to this great work that is unreliable that it is difficult to arrive at the exact truth. That the animal and the vegetable world, as they now exist, will sometime in the far-distant future be buried out of sight is very probable. The carboniferous period discloses many varieties of fauna now extinct. The matchless trees of California are passing away and the scythe of time is continually sweeping down many species of the vegetable and the insect world never to rise again. If we could reach far down into the secrets of nature it is not improbable that we might discover that many species of the human race have been subjected to an inevitable law which carries in its train slowly but surely the elements of desolation. In the far distant future Time with his hoary locks will leave naught of American greatness but the decayed ruins of our rural villages and the scattered fragments of granite and marble of which our splendid cities are composed.

When this great cemetery was first discovered it was noticed that the summit was depressed, so as to give it a dish-shaped appearance. On that account Tomlinson was led to further continue the search; he commenced at the top and sunk a shaft. Nothing was discovered until he came within thirty feet of the lower vault where he found a single skeleton so much decayed that its bones were soon reduced to ashes. This skeleton was to all appearance buried alone. A vault of wood had been prepared the same as for those who were placed below. The top of this vault had been covered with stones, and when the timber decayed they fell down upon the skeleton, piling rocks and earth upon it. The shock disturbed the soil to the apex of the mound and gave it the dish-like appearance.

It has been an enigma to antiquarians why such a great and expensive monument requiring the toil of many thousand lifetimes should be constructed to cover the remains of but three human beings. That these personages occupied a very prominent place in the hearts of their people is beyond a doubt. Some suppose the males were priests representing a high position in their pagan religion; and holding in their hand the eternal destiny of the people and through fear of their dreadful gods whose thunders rent the sky and whose lightnings shivered the knarled oak, they moved the multitude with uncommon powers of oratory and led them where they willed. There is scarcely a nation but has its leaders. The Indian tribes are powerfully effected by the magnetism of many of their orators

in whom rests the power to lead them to deeds of
mercy or deeds of cruelty. A writer who listened
for two hours to a speech delivered by Red Jacket
on a great occasion says, "He moved his auditors
as the wind in its rushing majesty moves the sublime magnificence of ten thousand forests." Oratory frequently dethrones the reason, not only
among the pagans but even among civilized people,
like that of Peter the Hermit, when he rode through
the eastern cities bareheaded and barefooted,
preaching the deliverance of the Holy Land and
inspiring the Byzantine warriors to terrible deeds
of cruelty. That the laying of the foundation of
this great mound at Grave Creek was done by some
sort of inspiration uncommon on ordinary occasions,
is very probable; and that the persons there entombed occupied some priestly or kingly position,
is beyond a doubt.

Some suppose that slavery existed amongst the
Mound-Builders, and that the vast labor performed
in building such great works was done by slaves
who were either born as such or were prisoners
taken in war. But from the appearance of the
relics found and the expensive mode of burying
their dead it is very probable that they were governed mostly by their priests, though they might
have had other powerful rulers or chieftains who
united with them in riveting the chains of servitude.
This great tumulus and other gigantic works built
with so much labor is satisfactory evidence that
the masses enjoyed not the blessings of liberty.
It looks very improbable that they would construct

such a great mausoleum over three individuals merely to perpetuate their memories. It is beyond a doubt that in the hey day of their existence they were a happy people overflowing with the blessings of prosperity and multiplying and replenishing the earth.

The skeletons and copper bracelets, above referred to were not the only relics found in the vault in the midst of this great mound. More than three thousand beads of curious and artistic workmanship, together with two hundred and fifty blocks of mica and large quantities of flint and copper implements were disclosed. But one of the most interesting relics ever found in any mound in this country was a block or flat stone containing three lines of unknown characters engraved upon it. This stone lay near the heads of the skeletons, and it was thought by Messrs. Tomlinson and Schoolcraft that it proved a written language and

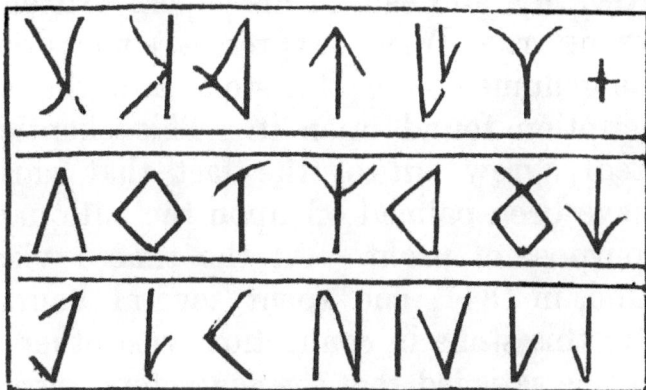

Fig. 13.—Characters engraved upon this block.

was designed to be read by the inhabitants of the spirit world. American archæologists have been

interested whether these ancient inhabitants of the·
long ago were possessed of a written language.
Many nations—formerly supposed even far in advance of the Mound-Builders—had not attained to
the construction of an alphabet.   So extensive was
the interest taken in this stone and the singular
characters engraved upon it, that Schoolcraft sub-
mitted it to many of the most learned professors
in Europe, among whom was Professor Rafn, of
Copenhagen, and M. Jomal, of Paris.   These distinguished linguists, after giving the subject a
thorough investigation, came to the conclusion that
the stone contained twenty-two alphabetical characters which they were unable to interpret.   Others
less learned than the latter gentlemen claimed
they were kind of mongrel Hebrew characters and
attempted to give a full translation thus rendered:
"Thy orders are laws."   "Thou shinest in thy
impetuous clan, and rapid as the chamois."

The Rev. Mr. McLean, in his papers written for
the STAR OF THE WEST, expresses some doubts
about the genuineness of this stone and the mystical inscription found upon it.   This skepticism
undoubtedly grew out of the fact that various
frauds have been palmed off upon the antiquarian
for the purpose of profit.   At the time I visited
this mound, in 1841, and spent several hours in
examining this stone in connection with other relics, I became satisfied that the characters engraved
upon it were done by the people represented by
the skeletons found there ; and I was also inclined
to the opinion that they had a written language.

Enough has been discovered engraved upon stone to show they had characters to convey their ideas. If they had a written language it is very improbable that it was always engraved on stone. They had the bark of trees and, very likely, they understood the art of tanning leather which they might have used for writing and printing their language. If they used such perishable materials they must of necessity have been obliterated centuries ago.

These ancient people must necessarily have had very extensive business relations with each other. We find them traversing the length and breadth of this great continent collecting and manufacturing materials for warlike and domestic purposes; we find their marks on the shores of Lake Superior in collecting copper on a large and gigantic scale, and then in distributing it throughout the length and breadth of the American Continent; then again in the mountain regions of North Carolina we find extensive works where they collected and mined for mica, an article frequently referred to heretofore, which is always found in connection with the bones of their distinguished dead. This article has been found in every section of this country where the Mound-Builders have left their marks. The Conewango Valley, nearly a thousand miles from the mine, has disclosed numerous large blocks of that substance — the one in particular found by Dr. Everett of very large dimensions, which, if split into sections, would cover a great number of square yards of surface. Several other blocks have been found in the immediate vicinity differing much in size. The circumstance that mica was so extensively distributed in the vicinity of the Conewango River is another proof that that section was once densely inhabited by the Mound-Builders.

Now when we take into consideration the extensive operations carried on in so great a territory and the products distributed over a large portion of this great country, it shows the inevitable necessity of a written language, either in the form of symbols or in alphabetical characters.

About two miles above the city of Steubenville, on the West Virginia side of the Ohio River, is a large rock composed of sandstone literally covered with strange hieroglyphics expressive of something unknown to us. While examining this rock, in 1875, I noticed that these engravings were made a long time ago and that many of them are now nearly obliterated. The examination of this rock which took place during a gloomy day brought a thousand thoughts to the mind. I saw that Time had not only placed his hand upon the ancient inhabitants and their works, but has laid cities in ruins, crumbled thousands of generations of magnificent forests down to dust, laid bare the mountain tops, drained great lakes of their waters and made of them fertile fields, caused the mad waters of Niagara to hew great channels of fearful depths as they rolled over their rocky bed, brought continents from beneath the ocean waves and caused millions of animated beings who sported in great rivers and ocean tides to go out forever. Time, that great destroyer, has mingled the soil with

human dust and the scattered fragments of human
skulls.

> " Look on this broken arch, its ruined wall,
> Its chambers desolate, and portals foul;
> Yes, this was once ambition's airy hall,
> The dome of thought, the palace of the soul;
> Behold through each black lustre eyeless hole,
> The gay recess of wisdom and of wit,
> And passion's host that never brook'd control:
> Can all, saints, sage, or sophist ever writ,
> People this lonely tower, this tenement refit. "

CHAPTER VII.

MICA—ITS EXTENSIVE USE AND GENERAL CHARACTER-
ISTICS—WHERE OBTAINED—ETC.

During the time I was contributing a series of
articles on " Pre-historic Man " to the RANDOLPH
REGISTER, when I had occasion to frequently refer
to mica — as I regard that mineral the most important in tracing the ancient Mound-Builders — I
was frequently asked the question, what is the
nature of mica and where is it found, &c.   I have
also been asked to give some geological facts with
regard to the different changes known to have taken
place upon the earth and the probable time man
appeared.

As I am desirous to touch upon all points which
will have a tendency to interest and instruct the
reader, I shall devote the present chapter to a brief
explanation of mica and some of the great geological changes that have been developing for millions
of years.

Mica is always found in connection with rock,
such as granite, gneiss and mica slate.   It occurs
in every section of the country where granite rock
is found.   It is capable of being cleaved into elastic

plates which are more or less transparent, and is frequently used in place of glass for lanterns and doors of stoves. Notwithstanding this substance is always found more or less in granite rock, there are but few sections in the United States where it is found in large blocks, such as are disclosed in the great mausolea erected over those ancient remains.

The reason why such large quantities of this substance should have been distributed so extensively and brought from so great a distance has been one of the greatest puzzles to archæologists. Some have claimed — which is undoubtedly correct — that they built wooden structures and occupied cities and villages, with windows in their dwellings, composed of transparent plates of mica, which would readily disclose to them the light of day. The locality where the village of Circleville and several other villages in Ohio now stand, shows evidence that large cities have been erected there. I have been informed by several persons familiar with that locality that numerous pieces of mica have been found in thin plates which, to appearance, had been used for windows. As mica was regarded with so much favor by those ancient people, it is not unreasonable to suppose it was used to some extent to decorate the ladies when they went into the mazes of the dance, and again when they made an afternoon call; then it would make an appropriate subject for gossip, as it would be contrary to every known law governing the female sex, if they failed to vie with each other in regard to the beauty and the shining qualities of the jewelry used to decorate their beautiful forms.

We have some evidence that the Aztecs used the mineral isinglass to some extent; and, notwithstanding the numerous theories with regard to the "Toltec migration," it is not improbable that the Mound-Builders were in some way merged into the magnificent Empire of the Montezumas.

That the Mound-Builders were controlled by kings, priests and royal personages is attested by their constructing such magnificent tombs to perpetuate the memory of the few; and it is reasonable to suppose that many of the relics buried with the dead were regarded as the insignia of majesty.

Montezuma, when he met the conqueror of his powerful empire, Fernando Cortez, appeared with a numerous and noble attendance. Three nobles preceded holding in their hands rods of gold to advertise the presence of their mighty sovereign. The great king came richly clad in a litter covered with plates of gold interspersed with plates of mica, which was borne on the shoulders of four nobles, under the shade of a parasol composed of green feathers, and decorated with gold and gems of dazzling splendor. He was accompanied by two hundred lords dressed in a style far more splendid than that of the nobles; but as they walked by his side their feet were made bare to show their humble reverence to their royal king.

It is not improbable that the locality of Grave Creek was once the seat of a great empire, and that the remains found enclosed in that expensive

sepulchre were the forerunners of the last of the
Montezumas, that great and powerful monarch
whose empire once so glorious fell by the treachery
of a christian conqueror.

The circumstance of such an amount of mica
being found in those great tombs makes it very
apparent that it was used to decorate their kings
on a scale very different and far more ancient than
that presented by the last of the Montezumas.

There never has been a people known in history
but have had some means of decorating their persons. Beads and feathers are the pride of Pagan
nations, but among the civilized, golden jewelry is
in use and made in various fantastic shapes.

As mica has occupied such a conspicuous place
in the unknown history of those ancient people,
and as we find it covering the whole of a skeleton
in many instances, it must have held an uncommon
importance and seems to be one of the most prominent keys to unlock the door to the dark and
mysterious tomb that leads to the history of a people whose tracks upon the sands of time are fast
fading forever. Notwithstanding some may differ,
it is very probable that mica, as frequently observed,
was in some way regarded with religious veneration, else why should such labor in procuring
and transporting it have been performed? The
mica usually found in these mounds is in a crude
state and has never been wrought into any purpose
for domestic use or for decorating their persons.

That which we find scattered in the vicinity of
their great works was frequently wrought into

ornaments showing a large amount of mechanical skill and artistic design.

The god of the ancient Mexicans was the sun, together with those brilliant luminaries that glisten in the blue vault of heaven. As mica is less transparent than glass and has a surface so finely polished as to defy the most ingenious efforts of art, it readily dazzles the rays of the sun as they are thrown upon it. Now it seems not inconsistent to believe — if the Mound-Builders worshipped the sun and the heavenly bodies as did the ancient Mexicans — that they regarded it as having some important influence with their great and glorious object of worship, and that when they arose in the eternal world, clothed in the garb of immortality, they would be ushered into a heaven made transcendently brilliant by thousands of plates of that substance which would distribute the rays of the sun with dazzling splendor.

The thousands of beads disclosed in the Grave Creek mound has also led to much discussion. Why and for what purpose were more than three thousand beads buried with those two persons, is a question, and is considered a great enigma. Various opinions have been maintained very wide apart. The more attention we pay to this subject the more plainly it appears that those ancient people taught the resurrection of the body. This great mound which was never designed to be disturbed by human hands contained articles of immense value to them. The blocks of mica found there would have loaded a wagon, and were transported hundreds of miles over rivers and mountain streams. The beds were mostly composed of marine shells brought from the ocean shores, hundreds of miles away. Bushels of stone implements which cost those people great labor to manufacture were also found. Now it is plain that a people with ability to construct such immense works would never have buried articles of so much value, unless they believed the dead would rise again in a bodily form and become legitimate heirs to the property buried with them.

The beads above referred to showed considerable mechanical skill in their manufacture. They were about the size of a twenty-five cent piece of silver. The holes for receiving the string were drilled with as much perfection as could be done with a modern lathe, and they had the appearance of having been turned in a similar machine. The material of which most of these beads were composed is supposed to have been brought from the vicinity of the Gulf of Mexico, as sea shells of a similar character are numerous in that locality. These shells were transported, as generally believed, to the valley of Grave Creek and manufactured there, as large quantities of broken fragments have been found in that vicinity.

WERE BEADS AND MICA USED AS MONEY.

Many suppose that beads with those people were a kind of circulating medium among the inhabitants and used as money. That the persons buried in that great mound were regarded as wealthy is very probable. They willed to have their wealth buried with them, so that when they burst the bands of death they would meet their friends glittering with the riches of the tomb and sporting with loved ones upon the shores of immortality.

About five hundred of the beads found in that great mound were composed of a peculiar kind of clay, *hydrate* of *alumina*, such as is used for the manufacture of pipes and earthen-ware at the present time. Those beads have the appearance of having been made in a mould and then subjected to a sufficient degree of heat to render them hard. If the theory entertained by many be correct, it is very probable that they placed a different value upon the numerous kinds of beads that were in circulation. It is also apparent that they had a circulating medium of some description, for we find them engaged, as already stated, to a great extent in the copper mines of Lake Superior collecting copper in large quantities and converting it into implements of warfare and for domestic purposes, and then distributing it from the region of the great lake to the Gulf of Mexico. The mica mines of North Carolina were extensively worked and the material is found in every sepulchral mound, and frequently in great quantities. In the United States alone the mounds of this character will number nearly fifty thousand.

Then again the flint business was very extensively carried on. The material for making arrow heads, axes, spear points and the great variety of implements made of flint is confined to but few

localities; but when we believe there is scarcely
an acre of tillable land in the United States but
has or did contain arrow points once used by those
ancient inhabitants — and many acres have disclosed them by scores and even hundreds — we can
form some idea of the extent of the flint and arrow
business. The doctrine of mine and thine is a part
of human nature, and is and always has been rec-
ognized by every nation. The Southern people
years ago felt its influence, notwithstanding they
bought and sold bodies of men and women and
plunged them into a slavery dark, dismal and hope-
less; but this strong element of human nature
outrode the storm and triumphed over chains and
fetters. A very intelligent gentleman from the
South informed me some years ago, in a conversation upon the subject of slavery, that the great
secret of their success was an occasional week of
jubilee and the use of articles they could call their
own.

Communities have frequently been established
governed by high moral and religious principles,
that seemed to possess every element to make a
happy people; but on account of the position of
the property, being owned in common, they rarely
survive the life of more than one generation. If
all of the pagan nations of which we have any
knowledge had and circulated some kind of cur-
rency, it is very apparent that the ancient inhabitants who outnumbered the hosts of the Indian
races by millions and soared far above them in the
arts, were not without a circulating medium. It

is not established that they made use of beads, as several archæologists have suggested, hence something very different may have been used in great abundance, which has been unable to survive the scores of centuries long since passed away.

If mica and beads were a circulating medium with those unknown people, it is very likely that copper was also used to a great extent for the same purpose; for when we consider the great amount of labor required in obtaining this article, it must have possessed uncommon value. But the fact that mica was frequently deposited over the dead no doubt caused the value of that substance to be more particularly connected with their superstitions.

The scientific world, within the past few years, has been more than ordinarily excited with regard to the age of the earth and the advent of man and animals upon its surface. Geologists have dug the hills, explored the mountains even to the regions of perpetual snow, followed the tributary streams of great rivers to their source and thence to the ocean tides. The seven seals of the great geological book have been opened, disclosing in written characters upon its rocky leaves the physical condition of the earth for millions of years.

Notwithstanding the great amount of labor bestowed upon that interesting subject, it is still in its infancy. Various theories have been adopted and accepted for a time which have now been abandoned, on account of new discoveries. Others which have been received with popular favor will also fall to the ground. The carboniferous doctrine

with regard to the formation of coal from the decomposition of vegetable matter is utterly fallacious
and will be abandoned when petroleum is proved
to be the progenitor of coal and the great agent of
the carboniferous period.

Geologists have endeavored to give us the first
dawn of animal life upon the earth from the testimony of rocks formed millions of years ago in
the silurian waters, where the fossils of the first
forms of life are found. They have divided the
different changes which have taken place on this
globe into different periods of time, which are
seven in number. The first is the azoic age, an
age devoid of life. The second is an age when
mollusks, corals and trilobites abounded in the
oceans before the continents were raised from
beneath the mighty waters. The third period disclosed numerous kinds of shells, worms and fishes.
The fourth period discloses plants and shrubs of
various kinds together with amphibians and reptiles. The fifth age was an age of reptiles out-
numbering and exceeding in size any which had
heretofore made their appearance. The sixth was
an age of mammals of giant size together with
reptiles in great abundance occupying both land
and water. The seventh age was that in which
man was introduced upon the earth, who presents
an organization the most perfect of any in the
zoological scale. This organization is the terminating point in the development of the great class
of vertebrated animals ever in existence.

Man is emphatically the noblest work of nature

and the only animal possessing powers of intellect
by which he can look "from nature up to nature's
God." But notwithstanding the strength of his
intellectual faculties, the lower animals in the perfection of some subordinate powers exceed him.
The eagle can soar into the distant sky and there
discern his prey, aided by the remarkable development of his eye and the acuteness of his optic
nerve. The dog can follow the track of his master
after hours have passed away, on account of the
great size of his Sheiderian membrane and the
energy of its nervous powers. The horse, the ox,
the elephant, and various other animals are far his
superior in muscular strength and powers of en-
durance growing out of the great size of their
physical frames. Nature has been bountiful in
developing animal life and adapting it to different
conditions and to different climes; for the tribes
that roam the arctic shores and traverse the torrid
sands are fitted each for his own proper sphere of
life. But the perfection of the digestive functions
of man in his omnivorous character distinguishes
him from all the lower orders, whilst the perfection of his life is made manifest not more in his
intellectual and moral powers than in his capacity
for inhabiting different zones and different climates.
Although this may be attributed in a great measure to the salutary precautions which his superior
intelligence enables him to take, yet, in a great
degree it is owing to the vigor of his constitution
and the energy of his vital principle. Thus, while
the wild gazelle browses upon the tenderest plant

and the vulture lives but on the carnage of the battle field, man thrives both on the products of the animal and the vegetable kingdom; and while the polar bear is confined, nay, chained to his own ice-built rock and the giraffe to his scorching desert, man dwells at will amid the eternal snows of Siberia and the breezeless sands of Ethiopia.

It is evident that man appeared upon the earth in connection with mammals, but not at the time when the lowest forms began to have an existence. The scale through which animal life has been introduced has been upwards. In latter times there has been much controversy in regard to the time man made his first appearance upon this planet. Prominent geologists place it from fifty to two hundred thousand years — the latter time being the opinion of Prof. Wm. Denton and also, we believe, that of the late Sir Charles Lyell, the popular English writer on that subject. In the " Encyclopedia Britanica, " under the head of ' America, ' reference is made to human remains being found in the conglomerate contained in a Florida reef which Agassiz claims to have been buried ten thousand years; also two remains found in the delta by Dr. Dowler, near New Orleans, beneath many buried forests which he estimates to have lain for fifty thousand years.

That man has existed on this continent for a long period of time is very apparent; but when we attempt to measure it by years, it is mere conjecture. That great animals were developed here long before the Eastern Continent was above the

waters, is now regarded an established fact.   Prof.
James D. Butler, LL. D., in one of the most interesting lectures that has ever met our attention,
delivered before the State Historical Society of
Wisconsin, on "Pre-historic Wisconsin," says:
"Whether America was falsely styled the New
World seemed, till lately, a problem of impossible
solution." But it is now confessed that Agassiz
following Sir William Logan's Laurentian azoic
researches has proved America to be "the first-born
among all the continents; her's the first dry land
lifted above the waters; her's the first shore washed
by the ocean that enveloped all the earth beside;
and while Europe was represented only by islands
rising here and there above the sea, America
already stretched an unbroken line of land from
Nova Scotia to the far west.

CHAPTER VIII.

GREAT STONE FORTIFICATION AT BOURNVILLE.

Having occupied considerable space with the mound located at Grave Creek, in the State of West Virginia, I will now pass over the Ohio River and examine some of the ancient works located in the great State that bears its name. I have stated in a former chapter that the fortifications and works for defense are more numerous in Western New York than in almost any other section of the United States, and constructed, no doubt, for immediate use and on a much smaller scale. The State of Ohio presents more remarkable works of that character than any other State in the Union.

One of the most interesting and singular fortifications designed for defense against a powerful enemy is located near the village of Bournville and about twelve miles from the city of Chillicothe. This work is situated on an eminence, some four hundred feet above the surrounding country, and is flanked on two sides by gentle streams that mingle their waters about a mile below the base of the hill. The summit is a broad plain and is noted for its remarkable fertility. It projects into the great valley formed by the junction of the two streams above mentioned. This fortification is constructed on a very different plan from any heretofore referred to. Instead of a ditch and a parapet composed of soil, this is composed entirely of stone, with a wall two miles and a quarter in length, and of sufficient width to inclose about

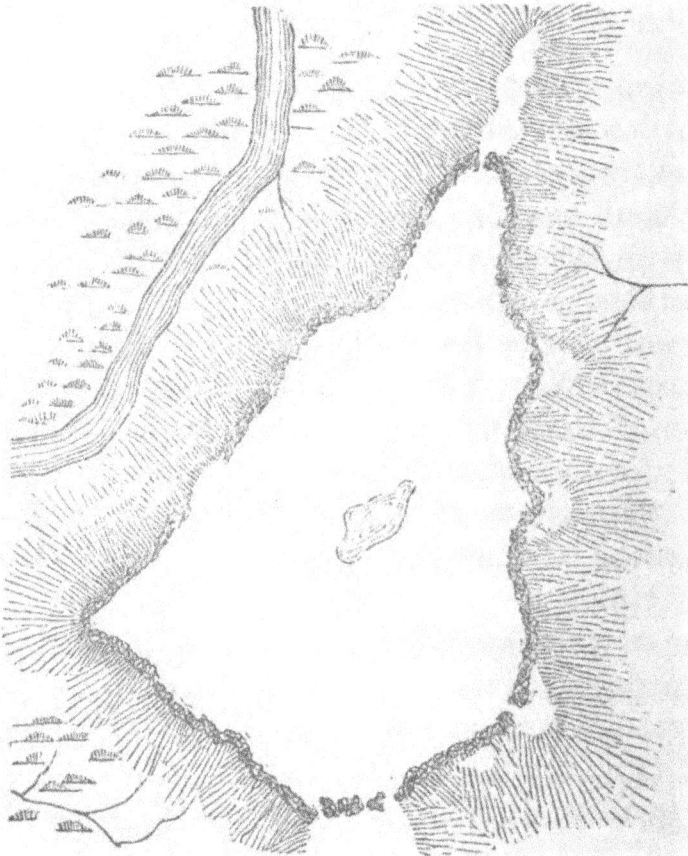

No. 14.—Great Walled Fortification near Bournville.

one hundred and forty acres of land. The wall varies somewhat in width, being on an average

more than fifteen feet. Along the western brow
of the hill, for eighty rods or more, it is from
thirty to fifty feet in width at the base. The height
of the wall, when first examined by the whites,
was about four feet and continuous, except at the
breaks which form the gateways five in number.
Originally there was still another gateway, but for
some reason best known to the builders it was
closed up. At the gateways, on each side of the
entrance, are immense piles of stone laid in the
shape of a pyramid with a wall extending out a
hundred feet or more. Within this great inclosure
are two large stone mounds which have been sub-
jeced to the action of fire and readily show the
effects of intense heat. The fires, from the general appearance, were burning for a long time and
used, no doubt, as signals of alarm in case of the
approach of an enemy. Adjacent to one of the
gateways are several pits excavated and cemented
at the bottom so as to hold water, which proved
a success, for even now, after the frosts and storms
of thousands of years, water remains in them the
entire year. The Rev. Mr. McLean in describing
this great fortification is of the opinion that the
wall was originally eight feet high or more. To
present an idea of the vast amount of stone used
in the construction of that ancient work Mr. McLean
says: "On the point of one of the gateways, after
a sufficient amount of stone had been removed to
form a line fence between two farms, their removal
was hardly perceptible."

This great wall was originally laid in a mechanical manner, but now it is mostly covered with gigantic trees whose roots have penetrated the soil and twisted amongst the stones, so as to dislodge them from their former position.

This great fortification to primitive people must have been impregnable. It overlooks a great expanse of country, formerly occupied by a large and extensive population, as various groups of works where buildings once stood and the dead were laid — but now in ashes — present themselves to view.

FORTIFICATION NEAR HAMILTON.

Another great and prominent military fortification is located about three miles from the present city of Hamilton, and at the distance of half a mile from the great river Miami. Its location is on the summit of a hill whose height is about two hundred and fifty feet above the base and the most elevated of any in that section of country. It is surrounded on all sides by precipitous bluffs which are almost inaccessible, except a narrow strip leading northward. The wall is composed of stone intermingled with clay taken from pits in the immediate vicinity, which can readily be seen at the present time. The wall will average about five feet in height, with a breadth of thirty feet at the base. This great wall built, no doubt, by the orders of some leading monarch who held his subjects in a mental and a physical bondage, incloses an area of more than sixteen acres. The wall is continuous, except where it is interrupted by four

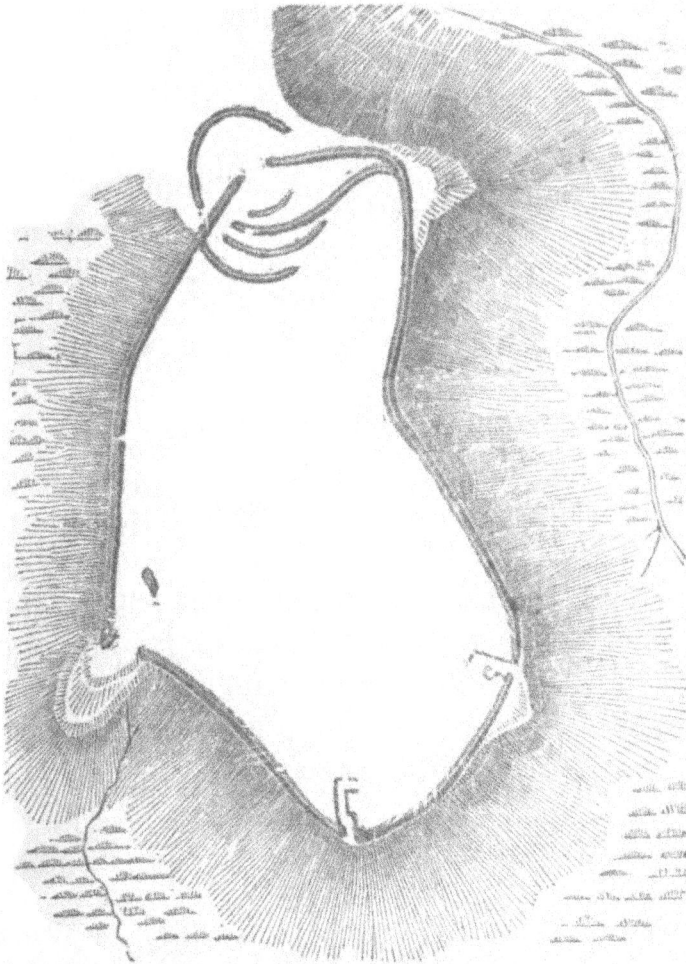

No. 15—Great Work at Hamilton.

gateways about twenty feet in width. The northern gateway is the only one easy of access from the surrounding territory. The wall at this point is built in the form of a crescent and is protected by four inner walls built in such a form as to bewilder an enemy should he succeed in forcing an entrance, and it gives a great advantage to the assailed, as they could shower them with deadly arrows ere they could rally for the conflict.

The centre of this enclosure is elevated or rounded up to the height of more than twenty feet above the base of the wall, so that from the location of the inner walls the party holding the fort could easily protect itself from the spring of an enemy many times its number. The summit of the elevation thus enclosed commands a view of the surrounding country for many miles distant on either side.

This great and extensive work should be sufficient to convince any reasonable mind that it is not the work of a people represented by our present Indian tribes. It is, in fact, as much of a mystery to them as it is to the most enlightened people now living. This fortification is constructed with as much skill as any ever made by the most prominent engineers who have graduated at West Point, or any other military school either in Europe or America.

FORT ANCIENT.

Another interesting work is located in Warren County, Ohio, thirty-three miles north-east of the city of Cincinnati, and has long been known as Fort Ancient. It has often been surveyed and visited by most of the archæologists in America and Europe. This great fortification is located on a hill about two hundred feet above the Little Miami. On the west, towards the river, the banks are steep and rugged. Two ravines made by the action of flowing streams originate near each other,

some distance to the eastward, and sweep around
the hill on either side of the fortification. The
position is very strong, being defended by these
two ravines and the precipitous banks formed by
the flowing waters of the Little Miami. The bank
forming the outer wall of this great work measures
about four miles in length and surrounds more
than one hundred acres. It is in several places
more than twenty feet in height by sixty in breadth
at the base. The soil composing this wall is clay
and loam and was taken from pits which are now
distinctly visible. There are more than seventy
gateways located at irregular distances along the
line. Messrs. Squire and Davis, authors of the
"Ancient Monuments of the Mississippi Valley,"
are of the opinion that they were places once
occupied by block houses or bastions composed of
timber long since decayed.

The embankment and earthworks connected
with this great fortification contain about six
million cubic feet of earth. The northern division
of this work, which would naturally be most exposed to the assaults of an enemy, is protected by
two extensive walls thrown across in the shape of
a crescent with the convex towards the north.
Within the enclosure are twenty-four reservoirs
fed by living springs that could never be exhausted
by an army of men thousands in number.

Near the exterior of the eastern wall are two
mounds of large size, which are surrounded with
a heavy embankment about four hundred feet in
length and join the main wall of the fortification.

As these mounds were constructed at the threshold of this gigantic structure, it is very apparent they were made to enclose the remains of some distinguished military hero who had played a conspicuous part in the theatre of ancient warfare and won a reputation blazoned with pomp and power which comes of a wholesale sacrifice of human life.

At numerous places occupied by the enclosure are found large stones which will weigh hundreds and even thousands of pounds, which for ages have been washed by the waters of the Miami and which must have required an incredible amount of labor to be carried up the precipitous bank to the height of more than two hundred feet.

The broad section of land bordering the river and its tributaries is the most fertile of any portion of the United States, and is uncommonly well adapted to the growth of Indian corn cultivated, no doubt, to a great extent by those ancient inhabitants. From the great lakes to the far distant South we find traces of its having been used for domestic purposes. Pestles and other implements for grinding and pounding the corn are disclosed. The fact of finding corn in a charred condition, caused by the action of fire, so that it will remain for ages without decay is evidence beyond a doubt. When we take into consideration the wealth of the soil through a large portion of Central Ohio, we get a clew to those great and extensive fortifications located in that vicinity. That this locality was once densely populated by a primitive people engaged in cultivating the soil is very apparent. But from some cause unknown to us the spoiler came waving the red flag of war dripping with human gore. The fertile fields, once so productive, became dreary wastes; famine touched them with her bony fingers and encircled them in her icy arms; and now after more than an hundred centuries have rolled through the devastating floods of time those great military structures, for magnitude and design, bid defiance to the most prominent fortifications ever constructed in modern times. But their great ruins now stand before us in all the nakedness of desolation, a melancholy witness of the destructiveness of man and the perishable feebleness of his greatest works.

ANCIENT WORKS AT NEWARK, INCLUDING MOUNDS AND FORTIFICATIONS.

In the town of Newark, Licking County, Ohio, is a very singular and interesting series of ancient works consisting of mounds, circular walls, ditches, avenues, &c., of such an extent as to cover about four square miles of surface. These works are so extensive and complicated that it is not in our power to give a correct panoramic view with human language. On the western side of these wonderful structures is a circular wall twenty-eight hundred and eighty feet in circumference and raised on an average about six feet in height. On the western section of this circle is a mound one hundred and seventy feet long and elevated about fourteen feet. This singular tumulus is so located

that it commands an interesting view of the territory under consideration, hence it has been named
"the observatory." Directly opposite this observatory, in a north-eastern direction, is a gateway
leading into an avenue three hundred feet long and
about sixty in width, with walls on either side
raised to the height of about four feet. This
avenue terminates in one of the eight gateways
of an octagon enclosing upwards of fifty acres.
At each corner of this eight-sided structure is
located a gateway directly in front of which, at the
distance of sixty feet within the enclosure, is a
mound five feet in height and from eighty to one
hundred feet in diameter.

In a southern direction from this octagon are
three parallel walls extending to the distance of
about two miles and located about twenty feet
apart. One of these lines, near the termination of
the others, curves so as to pass down a bluff where
a large amount of earth has been removed to build
an elevated road over the swampy ground at the
base of the terrace.

At the termination of the middle wall is a square
which encloses about twenty acres. Within this
enclosure is located seven mounds circular in form
and symmetrical in every part. South of this wall
is a gateway leading into a wide avenue which
enters the gateway of another enclosure, at the
distance of about nineteen hundred and fifty feet.
The last named work is elliptical in form, and its
diameters are respectively twelve hundred and
fifty and eleven hundred and fifty feet. The em-

bankment which encloses this remarkable work is about twelve feet in height and about fifty feet in width at the base, and has a ditch on the interior side thirty feet in width and about seven feet in depth. At the gateway, on either side, the embankment is elevated so that at the entrance it is sixteen feet. The width of the gateway is eighty feet, and at the ends of the main wall where the opening is left they make a gentle curve and terminate in a small mound about twenty-nine feet in height.

In the centre of this enclosure is one of the most singular effigies ever discovered; it is in the form of a bird with the wings expanded. The length of the body is one hundred and fifty-five feet; width, sixty-three feet, and the height from the base is about seven feet; each wing measured from where it unites with the body is about one hundred feet and about forty-five feet wide in the centre. The head of this remarkable effigy points directly towards the gateway of this singular enclosure. In the rear of this bird, at the distance of one hundred feet, is a semi-circular embankment two hundred feet in length.

At the north-eastern section of this great work is another embankment of larger dimensions than any heretofore mentioned; but as the diagram before us fails to give the figures, I am unable to describe it only in idea. Several other parallel walls are marked in the diagram. In the eastern section twenty circles are represented in the survey, averaging about two hundred feet in circumference, all having gateways opening towards the east. A few hundred feet from the last mentioned circles are eleven others which will average six hundred feet or more in circumference. These are entirely surrounded with a high wall but devoid of gateways, which is not usually the case.

This effigy of a bird found within those great walls has excited the thinking world with astonishment. Why and for what purpose it was constructed still remains a mystery. Mr. McLean regards such structures as being devoted to sacrifice, and as this, when excavated, was found to contain an altar makes it very probable that such was the case. This altar was composed of burned clay, and in size about twelve feet in length by ten in width. The top disclosed a large amount of calcined human bones together with elaborate carvings in stone, copper instruments, ornaments cut in mica, beads, pottery, etc., were disclosed.

As birds were frequently found in effigy, it was no doubt connected in some way with their religious worship; and as they reverenced the sun, moon and stars, and as the bird is the only animal constructed with power to soar into the distant sky, they might have regarded it as a messenger to bring tidings from the gods whose home was in the starry heavens.

With regard to pre-historic times one thing is demonstrated beyond a question, and this is that monstrous animals once existed, many of which were contemporary with man. Serpents crawled upon the earth, compared with which the great anaconda is but a pigmy. Great alligators, fifty feet in length, had their home in the turbid waters of the rivers, and the *megalosaurus* and *megatherium* have left their bones to demonstrate their giant size and former existence. Great birds once existed whose bones exceed those of the ox in size, and with expanded wings could ride the boisterous waves of the hurricane.

Persons who visited the Centennial at Philadelphia, in the year 1876, will recollect the skeleton of a gigantic bird, whose bony frame and neck towered to the height of nearly fifteen feet—a demonstration that it once existed. Now it is not unreasonable to believe that such a bird lived with the Mound Builders, and went down with them, perhaps when the whole solar system was plunged into an icy wave.

Among the collections published by the "State Historical Society," of Wisconsin, is an interesting paper by the Hon. John T. Kingston, which refers to a tradition believed in by the Indians. I will quote his words:

"Just above the city of Alton, Ill., high upon the face of the cliff fronting the Mississippi, there was the picture of a bird standing erect with wings extended; it was represented as having horns like the deer or elk. In height it was about ten feet, and from tip to tip of the wings about twenty feet. Forty years ago the color was bright and very distinct, but in late years owing to the encroachment of the city, and continued smoke of the lime kilns in the immediate vicinity, it has be-

come very indistinct.   The legend of the Piasau
as related by the Indians is as follows : ''

" ' Many moons ago, before the white man came,
this enormous bird, the Piasau, suddenly appeared
in that country.   It extended its flight over a great
many miles of surrounding prairie.   It was so
large and strrong it would carry off both men and
women; even the deer were made its prey.   Its.
home was in a cave of the cliff mentioned above ;
its perch morning and evening was on a point of
the cliff, immediately above where the painting
was made.   There it would remain perched until
the sun was fairly risen, and then soar away in
search of prey.   Almost every day one of the tribe
would disappear, and too well the Indians knew
his fate.   The word Piasau told it all.   The victim was carried to the cave and in a short time the
bones only were left to tell the tale.   So great
was the alarm of the Indians in the neighborhood
that they fled many miles away, but they could
not escape the flight of the dreaded Piasau.   Every
stratagem that they could invent was resorted to
for its capture and destruction, but all without
avail.   At length the chief, an old man, fasted
many days and nights and prayed to the great
spirit to save his people from destruction.   One
night the great spirit appeared to him in a dream
and told him he must sacrifice himself for his nation; said that in the morning before it was light
he must take his station on the highest point of
the cliff where the Piasau made its usual appearance at that early hour; that he must place twelve

of his bravest warriors in ambush close by, with bows and poisoned arrows, and that when the Piasau discovered and darted down they must let fly their arrows and if possible kill or wound him, and this they could do if their hearts were brave. He accordingly choose out twelve of his bravest warriors and placed them in ambush as directed, and then took his station on the top of the cliff, covered his head and then began to sing his death song. Just as the sun was seen rising in the east the Piasau appeared; soared up, and circling around, high up in the heavens, made the fatal swoop for the chief, but just before he struck him with his talons the concealed Indians let fly their arrows and the Piasau fell dead, pierced through the heart. The chief was saved alive and his people were saved from destruction.'"

The cliff above referred to is two hundred feet above the Mississippi River. The banks are nearly perpendicular and as you view the river from those dizzy heights it not only presents a grand but a fearful appearance. The cave is located about midway up this assent, and when first examined by the white man it was literally strewed with dead men's bones, and the bones of various animals mingled together in great confusion. How these bones became deposited there is a mystery. It is true there was once an ancient people who dwelt in caves but never under circumstances like this. This legend of the Piasau, although it presents much of the marvelous, is worthy of investigation. The great skeleton at the Centennial proves the former existence of a monstrous bird. Some years ago I saw in several papers a statement of the finding of some singular bones of large size and hollow cylinders. The writer, perhaps not being familiar with the osseous formation of the bones of birds never suspected that they belonged to a winged animal, but to-day I will say with the immortal bard that "There are more things in heaven and earth than is dreamed of in our philosophy."

If such a bird lived and flourished in this country, why not find the bones more frequently, as the bones of the mastodon are often found. The mastodon was a land animal of large proportions and frequented swampy places where he often lost his life, and in a comparative short sime became completely enveloped in a soil containing various elements of preservation. The Piasau (if he ever existed) would seek the mountain regions and the dry land where he would leave his bones, which, unless in some peculiar localities, would long ago have crumbled and mingled with the soil.

Several species of monstrous birds were formerly known to man, which have become extinct, within the last hundred years. Among them was one that somewhat resembled the ostrich, and lived in the mountain regions of New Zealand, and frequently reached the height of twelve feet. Another called the Epiornis, whose size was still greater and whose remains are frequently found in Madagascar. This giant bird if he possessed a carniverous organization would have carried an

Indian or Mound Builder with ease. But those ancient people, no doubt, conquered and tamed the great beasts as a general rule, or they could never have lived so numerous, and occupied a territory reaching from the Atlantic to the Pacific Ocean and from beyond the great lakes of the north to the southernmost extent of the American continent.

DID THE ANCIENT PEOPLE TAME AND WORK THE GREAT AMERICAN ELEPHANT?

I have heretofore suggested that the ancient Mound Builders were contemporary with the mastodon and that in all probability they tamed and used that powerful beast to haul heavy burdens. As I stand almost alone, in relation to that theory, I will give my evidence for such a belief. It is a fact admitted by all familiar with pre-historic discoveries that the bones of the mastodon and those of the Mound Builders are found in the same localities, and in about the same state of preservation; also in and around their great works, stones are frequently discovered with animals engraved upon them which are supposed to represent that animal. The copper relic, formerly referred to, found on the Allegany River with the form of an elephant engraved upon it, represented in harness, first attracted my attention to that subject. If the ancient people in North America tamed that great beast it is very likely that the inhabitants of South America done the same thing.

In Herndon and Gibbons explorations of the valley of the Amazon, in South America, a singular animal engraved upon stone and sketched by Mr. Gibbon is represented and no doubt was designed for the Mastodon, though it is devoid of tusks. Engravings of a similar character have been found in several mounds in different sections engraved upon bone.

Fig. 15.—Engraving on bone found in a Mound.

The fact that the mastodon was contemporary with man needs no argument for that is admitted by all antiquarians. Some may argue that on account of the great strength and power possessed by that huge animal it would be impossible for man occupying the position of the Mound Builders to bring him under subjection. If he belonged to the carnivorous class, it in all probability would

have been impossible to have tamed him, but as he
represents a far different order and was possessed
of a mild disposition like that of the African
elephant he could have been tamed when young
and brought under subjection with as little trouble
as is required for the ox or horse. The inhabitants of Africa, who no doubt stand lower in the
scale of development than the ancient Mound
Builders, use the elephant as a beast of burden.
There is scarcely a nation or people so low in
mental cultivation and the arts but resort to some
of the lower animals to transport their heavy burdens or carry them upon their backs. When we
consider the magnificent works built by these
ancient people it looks impossible that they could
have been built by no other than human labor.
The great mound at Cahokia, Illinois, is estimated
to cover twenty millions of cubic feet of earth,
which was all brought from a distance. Now, it
would take one thousand men nearly twenty years
to perform the labor which was bestowed upon
the building of that one tumulus, and when we
consider that that is but one of about sixty other
structures by which it is surrounded, one thousand
men could not have performed the great labor in
the days and years allotted to human life. If one
thousand men were employed upon those great
works for forty or fifty years it would surely have
taken nearly twice that number to have supplied
them with food, clothing, fuel and other necessa-
ries during that long period of time, and then
again we must suppose a numerous train composed

of women and children and feeble persons continually followed them which had to be fed, clothed and sustained.

In numerous places where these great works are located we find roads or graded ways leading from the valley or banks of rivers up various terraces which connect with the works at different points. One such occurs at Marietta and one at Piqua and another at Piketon which is constructed on such a gigantic scale that if they worked the mastodon they could have doubled their teams with but little trouble.

Fig. 16.—View of the graded road near Piketon.

The State of Ohio contains at the present time a large population and so far as the mechanical and agricultural arts are concerned they are far in advance of the builders of those ancient structures.

But, notwithstanding, their superior knowledge, I will venture to say that without the aid of iron implements and the use of teams to plow the land and transport the products to different sections, it would be impossible for them to be sustained and to construct such great works as we find scattered throughout the State. Twelve thousand different works have already been discovered within its boundaries and many others formerly located near the banks of rivers have been washed away by floods. When we examine this subject in its various bearings, and consider that we find graded ways and macademized roads in many places, which, now show the pressure of heavy loads having been borne upon them, it seems evident that they had the use of some powerful animals which they trained and educated. At Marietta, near a graded road, occurs a singular mound which has four graded ways reaching its summit. This work is represented in Figure 17.

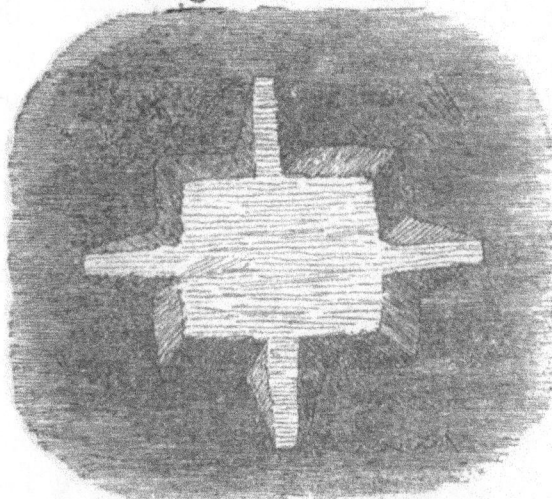

Fig. 17.—Ancient work at Marietta.

TRUNCATED MOUNDS.

The largest and most imposing mounds found in Ohio and other States, are truncated pyramids with graded avenues, reaching their summits; these are classified as "Temple Mounds." The summits of this class are supposed to have been crowned with temples made of wood, and used for religious purposes. The great mound located at Cahokia, Ill., belongs to this class, and is the greatest ever discovered among the works of the ancient inhabitants. This tumulus was built in a different form from that class called sepulchral mounds; those are always a true circle, and were repositories of the dead. The form of this great work was that of a parallelogram. The sides at the base were respectively forty-three rods, and thirty-five rods, and occupied an area of six acres. On the south side was a broad terrace extending about eighteen rods, which was reached by a graded way that extended to the summit, which was ninety feet from the base and many feet above the highest steeples of our country churches. The summit of this mound is a level platform, that comprises about two acres, and contains a sepulchral mound about ten feet in height, and when excavated was found to contain a large amount of human bones, together with pottery and various implements of stone. Why this sepulchral mound should have been built upon the summit of this great work is not understood. Some suppose that the builders of that great mausoleum were a more recent people, perhaps the descendents of the conquerers of the race who built such gigantic structures.

These temple mounds were in all probability devoted to sacrifice. Evidence is abundant that fires have been continued upon them for a long time, for they invariably contain pits filled with coal and ashes interspersed with calcined human bones. These mounds bear a near resemblance to the sacrifical mounds found in Mexico and it is an established fact that the Mexicans were addicted to human sacrifice, and performed the ceremony upon the top of those temples by tearing out the hearts of the victims and scattering their blood upon the altar to appease the wrath of their dreadful gods. Cortez, notwithstanding he was far below them in honor and integrity, done a noble deed in abolishing this terrible practice of human sacrifice, which they were taught to believe from their childhood, and regarded a divine institution.

The more I investigate this subject the more I am inclined to believe that the Mexicans are legitimate descendants of the ancient Mound Builders. The forms of their temples are similar, they practiced human sacrifice upon the summits of their great temples, they used implements of warfare very similar to those which are found scattered so extensively among the ruins of the lost race. The Mexicans worshiped the sun and we have abundant evidence to believe such was the practice of that people who built so many great monuments which we find in this section of country. Engravings are frequently found representing the sun, moon and stars, which were with them objects of worship.

The Mexicans had a tradition with regard to the history of their people. When Cortez demanded of Montezuma to own allegiance to the king of Spain, he was wearing the crown of the king of Mexico, but was forced to pay tribute to a distant monarch. At the stern demand of Cortez he assembled his nobles and informed them of the orders of the conqueror, and in a solemn and affecting manner, with streaming tears he spoke as follows: "I speak as the Gods direct me. You know our diviners have told us that as other nations possessed this land before our ancestors came hither, and as our fathers supplanted their predecessors, so must a new race overcome and supplant ours. That hour has come, the scepter has passed from my hands, the crown which my fathers wore and which you placed upon my brow, must be laid at the feet of the stranger who has appeared among us, and who now requires me to surrender my power and your allegiance to the will and the service of the king of the east."

This great temple located at Cahokia, which resembled those of Mexico was undoubtedly devoted to human sacrifice, and various other religious ceremonies. From its uncommon size and being located in the vicinity of about sixty others —but of smaller dimensions—it was evidently regarded a place of great importance; a kind of Mecca where the religious pilgrims were wont to meet at stated periods for religious worship, and to do homage to the priests, and bow before the invisible gods, while the warm blood of the sacrificed victims to a terrible superstition was streaming upon the altar.

If the Mound Builders were merged into the great and magnificent empire of Mexico, through the Toltec migration, it would seem that some terrible cause produced the result. They were the legitimate heirs to a vast amount of territory. The great prairies of the west, the most fertile lands in the world were theirs, the alluvial lands of the great rivers was the home of thousands. Their works which lay scattered in almost every state, and which cost immense labor to construct, were left behind   Millions of acres of garden beds and cultivated fields were left for forests to grow and perish upon their surface. In the vicinity of the greatest works, the climate is now the most healthy and agreeable of any in America. If there is an "earthly paradise," that paradise was theirs.

Now if the theory entertained by sevaral archœologists, that those ancient Mound Builders left such a great country and so many great works is true, how did it happen and how brought about ?   I have formerly been of the opinion that wars, idolitry and superstition were the leading cause of their destruction. But it seems almost impossible that such a numerous people could become totally annihilated.

It is a fact known to geologists that great changes have taken place upon this earth of ours at different periods of time. There have been several glacial periods in both the northern and southern hemispheres, at which time countries in our latitude were clothed with a mass of ice. Now it appears not unreasonable to suppose that great changes are periodical, and at times the earth is thrown into a cold wave, which would drive all organized beings to warmer climes, or cause their destruction; such a state of things might have happened to the ancient people which drove them from their property and their homes to the vale of Mexico, where they have retained their religion, their gods, and modes of sacrifice upon altars resembling those which they left behind.

That a sudden and instantaneous change has at least once taken place is not improbable. The time once was when great mammoths inhabited Siberia and many northern climes, and to-day their bones are scattered in great confusion over a large amount of territory. Various plants are found frozen in great masses of ice, which belong and grow in semi-tropical countries.

In a paper written on the "probable future of the human race" by Alphonse de Candolle, and translated from the French for the Smithsonian Institution, occurs the following paragraph:

"Finally who can forsee what may happen to our entire solar system? It is moving with great rapidity in a certain direction. Perhaps sometime it may come into some part of the universe much warmer or colder than the space it has passed through for several millions of years. The sun may also change. Events such as these may destroy not only man, but all the organized beings of our globe."

The cause of this wonderful state of things is so obscure that writers have but little to say upon that subject. All admit, however, that it must have grown out of some periodical change which must take place from necessity, in the revolution of the sun and solar system, around some great center—where it was thrown into some unknown conditions, which produced the intense cold at the time of the glacial period—and perhaps in a less degree in more subsequent times. Now as such changes take place at different periods, it seems they must be *gradual* and require long ages to be brought about ; so that man if placed in northern regions would be apprised of their approach and flee to more southern lattitudes. But there is a fact that seems to contradict the theory of gradual change, and that is that the remains of the mastodon found in Siberia have the appearance that the animals were destroyed about the same time. Their bones are in a similar state of preservation, and have been found in such quantities that they have become an article of commerce. Ship loads have been carried to different parts of Europe, and also to America, and there wrought into articles of various kinds which require the purest of ivory. If the change was gradual it seems impossible to construct a theory adequate to solve this important question.

The mastodon, or ancient elephant, during the present century, has been discovered in icebergs, clothed with flesh, and in a perfect state of preservation, which would indicate that he fell a victim to a sudden and in-

stantaneous change.   As this northern elephant was
clothed with hair, it would appear that he was not ex-
actly adapted to a tropical country, but to a country
which could compare, perhaps with France, or the
southern part of Germany, which theory I believe is
now generally adopted.   But as France, and in fact all
of Europe, was once exposed to the glacial epochs, it
would seem that all animal life in all those sections
must have been destroyed at that time.

It is difficult, however, to arrive at any definite conclusion whether these huge monsters appeared before or
subsequent to the glacial epoch   Some are of the opinion that not only the mastodon, but that man, together
with various extinct animals which were contemporary with him, saw the great masses of ice as they floated southward and transported large quantities of earth
and stones from higher to lower lattitudes, but the
time which has elapsed since these wonderful changes
took place is buried far beyond the feeble intellect of
man.

Fig. 18—Ideal view of ancient man and animals that flourished with him.

WORKS ON THE MUSKINGUM RIVER AND OTHER LO-
CALITIES.

The localities where the greatest works of these ancient people were situated were well selected, and possess the necessary requisites for building a commercial
and attractive town or city; hence when the white
man came he was attracted to these localities and built
upon their ruins. In Ohio the cities of Cincinnati,
Dayton, Xenia, Norwalk, Marietta, and in fact nearly
every prominent city in the state is located on ground
once occupied by a people whose great works and decaying skeletons are all that remain to tell of their former existence.

At Marietta, where is located some of their greatest works, the facilities are natural and interesting. The great River Ohio is here met by the Muskingum which flows through one of the most fertile sections of the state, presenting scenery grand and beautiful, changing from broad and expansive intervals to precipitous hills and bluffs covered with nature's forests.

Above the junction of the Muskingum with the Ohio River, is at the present time a remarkable group of works, which at the time of the first settlement of the state, filled the people with wonder and astonishment. At that time the state of Ohio was mostly a wilderness, and the early settlers had but little knowledge of the amount of great works scattered throughout the country in almost every direction. In a few years after the settlement was made at Marietta, it became a place of much attraction, on account of such a great amount of ancient works located there.

As we ascend the Muskingum River the distance of about a mile, we are there presented with fortifications of amazing magnitude. These works are located on a broad and sandy plain, at the distance of about half a mile from the river, and about one mile above its junction with the Ohio. They are on the second terrace and about fifty feet above the border of the great stream. This site commands a splendid view of the meeting of the waters of the rivers; the one rising amidst the mountain regions of the north, and forming a great highway from near the northern lakes through Western New York to its junction with the

Monongahela. and then to the Gulf of Mexico. The
other traversing a country presenting the most enchanting scenery of any.in Ohio. From these works can
be seen the broad alluvial terraces bordering the Ohio
for miles in either direction.

Across the river to the distance of about two miles,
this broad interval is flanked by hills cut asunder by
flowing streams that tell of the work of ages. But in
connection with the beauty and grandeur of this location, it is very likely that very many of the works of
this ancient people were standing before the magnificent structures of Luxo. Thebes or Karnac had an existence; but those great works of elaborate magnificence have rose and fell; but of these ancient American monuments, many are standing to-day, in the un-
watered sand, in solitude and silence.

When the European first set his foot upon the soil
of Marietta it was covered with great trees, which always attain an extraordinary size in such localities,
growing out of the ancient cultivation and decaying
skeletons.

Almost every variety of works have been found here.
Sepulchral mounds were very numerous, together with
fortifications, temple mounds and graded ways, many
of which have been leveled down to give place for another people. Several square embankments can now
be seen which contain temple or truncated mounds.
Within the larger of these works are four mounds.
Three of which have graded ways that reach their
summits. The largest in this group is one hundred

and eighty feet long by one hundred and thirty-two in width. This has four avenues of ascent, located about midway upon each of its sides, all of which reach the summit. These avenues which were constructed for ascending the mound are twenty-five feet in width by sixty in length.

Another mound connected with this group presents a somewhat different appearance, but of smaller dimensions. This is about one hundred and fifty feet in diameter, and eighty feet in height. Three graded ways are connected with this tumulus, and reach the summit; one passes out to the north, one to the east, and one to the west. On the south side is a peculiar kind of terrace, or platform, which extends out from the body of the mound about fifty feet. This platform is supposed to have been occupied by orators who stood in that elevated position and descanted upon matters connected with their political jurisprudence, and their government and cares ; or perhaps it was occupied by priests, clothed in their sacred robes, teaching their dying fellow men the road that leads to the abode of the gods, whose throne is the sun, and whose eyes are twinkling stars which glisten in the heavens.

These remarkable works are regarded by archæologists as being very ancient, and were undoubtedly constructed hundreds of years before the foundations of many others of their imposing monuments were laid. It appears evident from the extent of these works and the circumstance of their presenting almost every variety known among the ruins so widely scattered, and being located on two navigable rivers, that a great variety of business must once have been transacted there. We can imagine that on some great occasion such as the death (in spite of the gods) of a prominent priest, a venerated monarch or military chieftain, a great throng gathered there to mourn for the departed, and to build a great mausoleum over his grave. The waters of our Conewango, long ages ago, may have carried down from its valley and the adjoining territory, to this prominent locality, scores of persons, perhaps to perform the sacred duties of religion, or to deliver a train of captives to be immolated upon some bloody altar dedicated to the gods.

GREAT TEMPLE MOUND AT SELTZERTOWN, MISSISSIPPI.

The temple mounds, which are quite numerous in the South, are rarely found in the North; these at Marietta are the most northerly of any that have been discovered. At Seltzertown, Miss., occurs a temple mound second in size to any found in America. This work covers about the same amount of ground as the great mound at Cahokia, Ill., but is forty feet in height. Its greatest length from east to west is six hundred feet, and four hundred feet in width, and placed in an exact posititton with reference to the cardinal points. At the base of the great work is a ditch passing entirely around it, which averages ten feet in depth. The summit of this work occupies about four acres of land, sufficient to contain a large concourse of people, who could overlook the surrounding territory when elevated to the height of forty feet. A graded way commences at some distance from the base, and gradually reaches the summit. Three conical mounds are placed upon this platform, one at the east, one at the west and one in the middle. The one at the west is truncated with a level summit thirty feet in diameter. The eastern mound is somewhat less in size and also has a level platform upon its summit.

These three truncated pyramids average nearly forty feet in length. Eight other mounds of small dimensions occur at different places, on the top of this great work, averaging about ten feet high. The north side of this mound was supported by a well built wall, composed of bricks, which had evidently been dried in the sun. These bricks were filled with grass, leaves, and rushes, which was mixed with clay or mortar, in order to hold it in a proper shape.

Dr. Dickeson the gentleman who first explored this mound, found a large amount of skeletons, and various specimens of pottery. The most singular find was several bricks, made in the form of tumuli, which seemed to have been deposited with remarkable care, and bore a perfect impression of human hands upon their surface.

SIMILAR TEMPLES IN MEXICO.

The great mounds above referred to located at Cahokia, Ill., and Seltzertown, Miss., bear a striking resemblance to the Mexican temples which were numerous at the time of the conquest. The City of Cholula was the central city of religion, the point highly favored with the gods, and was consecrated to their worship. A great temple mound was located here and in the immediate vicinity of forty thousand houses. On the summit of this truncated pyramid was erected a temple composed of wood and dedicated to the invisible divinities. So important was this holy temple in the estimation of the faithful, that pilgrims from far distant sections resorted here in great numbers to receive the blessings conferred by their holy religion, and to do honor to the devoted city.

When Cortez arrived at Cholula, on his way to the city of Mexico, the inhabitants, after consulting the priests, armed themselves for the destruction of the Spanish army; but notwithstanding the prayers of the priests and the promises of the gods, after the battle was fought, six thousand Mexicans lay dead upon the field. Previous to the commencement of this bloody battle, the priests tore from the arms of mothers ten little children to be offered upon the altar a sacrifice to the god of war; but neither the blood of little children spilled upon the altar, nor the sacred promises to the god of war, nor the intercession of priests, could stay the force of the artillery of the Spanish army which was brought to bear upon them.

The circumstance of the Mexican temples bearing such a close resemblance to those of the North, would indicate that there must have been some alliance between them. Those at the north, as mentioned heretofore, show the effects of fire which had been long continued. They also show altars which to all appearance had been devoted to sacrifice. The summits of the Mexican temples were crowned with wooden structures from forty to fifty feet high, where were placed the sacred immages of their presiding deities. The temples at the north show unmistakable evidence that such structures once were placed upon their summits.

At the time of the fall of the empire of Mexico they had a tradition which they preserved with religious veneration. They taught that their ancesters lived far away to the north in a country which they had occupied for a vast and unknown period of time, and from some calamity of nature they were driven south to avoid the intense cold that chilled the blood; that for more than one hundred years they continued with their descendants in a southerly direction residing in rude cabins, and planted corn and other seeds. At length they reached the vale of Mexico and there founded a glorious kingdom. They built up cities, elected a king, established wholesome laws, and practiced the working of gold and silver and the polishing of precious stones. We have no evidence that the ancient Mound Builders worked the precious metals to any extent in the north, but as Mexico is in the vicinity of gold and silver it is very likely they acquired the art after changing their location.

THE GREAT SYMBOLICAL MOUND LOCATED IN ADAMS

COUNTY, OHIO, REPRESENTING A SERPENT.

The most noted and singular effigy ever discovered in the United States is in Adams County, Ohio, and represents a serpent. It is located on a spur of land in the form of a cresent, which is elevated about one hundred and fifty feet above a rapid stream, which has poured its waters for ages against its base and formed a precipitous bluff. The summit of this hill is about ten rods wide by some sixty in length, on which is located the effigy of a serpent, so constructed as to conform to the shape of the land on which it is placed; its head is resting near the highest point; its body is about forty rods in length and winds back from the head in graceful undulations terminating in a tripple coil at the tail. The body of this figure is boldly defined, and is nearly six feet high by thirty feet through the base at the centre of the body. It is represented with open mouth apparently endeavoring to swallow some animal, the effigy of which is placed almost within its distended jaws.

To construct this symbolical work must have cost a great amount of labor, as the soil was mostly brought up a steep and rugged bluff. This work was no doubt nearly a third more extensive when completed than at present, as it has been exposed to the action of the elements for a long period of time, and has been covered with many generations of forest trees whose roots have penetrated the soil, and caused it to be leveled down by the actions of the storms of rain and snow which have beat upon it for centuries.

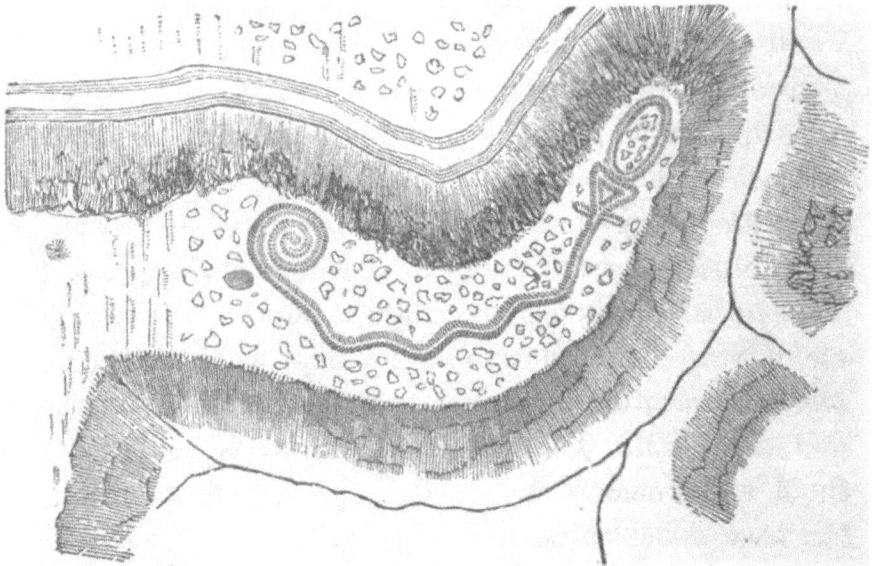

Fig. 19—The effigy of a serpent located on a hill in Adams County, Ohio.

Effigies of this character are frequently found, but never in connection with those of animals of a higher order. In the state of Wisconsin, where is located such a great amount of symbolical figures, none representing a serpent have been discovered. The effigies of serpents are nearly always found in connection with low, degraded animals. The object these primitive people had in view in constructing such vast and singular structures appears inexplicable, but as with nearly all their standing memorials, we must trace it to their superstitions. That those ancient people believed in another life beyond the grave is believed by all who have investigated the character of their works; and that they believed also in presiding deities is very probable. There is no animal in existence whose presence is so detestable as that of a crested snake—which delights to encircle both man and animals in its cold and deadly coils. Even savage nature untamed and untamable, shrinks with dread from the poisoned chalice of the viper's fangs.

Some have supposed that all of the symbolical mounds built by these ancient Mound Builders were calculated to represent some object of worship. This theory may be true so far as some of their effigies are concerned, but I have no confidence in the theory that a people so highly developed as the Mound Builders have shown themselves to be by their great works so artistically made, would worship one of the lowest and most degraded of reptiles. I am inclined to the opinion that the serpent with them was symbolical of a devil or infernal spirit, whose sparkling eyes would point to the slumbering fires within which would engulf them in everlasting pain and destruction, and that this great effigy was built with open mouth ready to devour its prey, to warn their fellow men to avoid the fatal snares of their hated enemy.

ALLIGATOR AT GRANVILLE.

Near the village of Granville is the effigy of an alligator, located upon the summit of a hill which is elevated about two hundred feet above the circumjacent territory. This hill commands a magnificent view of the country for many miles in extent. This symbolic structure is two hundred and fifty feet in length, and the breadth of the body near the head and shoulders is forty five feet, at which place it is elevated to the height of six feet. The legs and paws are respectively thirty six feet in length. Such is the nature of the soil of which this effigy is composed, that after being exposed to the actions of the frosts for unknown centuries, even the spread of the toes can be plainly indicated.

The superstructure of this work is composed of fine clay, which must have been brought from a great distance, for no indications of that substance have been found in the vicinity of the hill on which it is located. On exploring this effigy, in the center of the body a large space containing stones which had been subjected to the powerful action of fire was discovered. Why those ancient people should construct an effigy at so much cost of labor of one of the most contemptible animals that crawls upon the face of the earth is truly wonderful. The Rev. Mr. McLean is of the opinion that the altar of burnt stones points to its having once been devoted to human sacrifice by the action of fire. That might have been the case; if so, it bears no resemblance to the temple mounds in the United States and Mexico, which were undoubtedly used for that cruel and superstitious purpose.

Some have suggested that those symbolical structures, which represent such low and degraded animals, were devoted to the punishment and destruction of heretics.

It has been the custom with nearly all religions, to inflict upon unbelievers in the true faith the most excruciating tortures which their fiendish cruelty could devise, and degrade the heretic even below the vilest of animals. A prominent religious writer has said, (referring to heretics,) "They should have upon their funeral pyre such companions as vipers that would encircle them, and burn in the fiercest fire for months and even years, without being consumed."

The religious faith of the builders of such great and numerous works must have been strong and abiding, for they mostly point to their wars and superstitions. It is very likely that some of the most intellectual of the race threw off the shackles of ignorance by which they were bound, and denied the gods. If such was the case, we can readily conceive that the priests would become alarmed and consign their hapless victims to a most degrading and cruel death.

The crime of heresy with most religions is regarded the greatest, and demands the most terrible punishment. Such is the heinousness of the crime, in the estimation of the faithful, that the criminal is entitled to no sympathy from his fellow man. Even females, that

gentle sex whose very nature overflows with benevolence and sympathy, have looked with composure upon
a heretic chained to a stake and writhing in consum-
ing fire. The divine and holy instinct of the mother
has paralyzed, while her son alive was being burnt for
denying the gods; but a child, uncontaminated with
superstition, would recoil at the horrible scene. From
the abundant evidence presented in the works of the
ancient people that they were largely devoted to their
peculiar religion, and, no doubt, had severe laws binding its subjects to obey its precepts, that they should
inflict upon those who should disobey, painful and cruel tortures in the presence of an effigy representing
some infernal spirit.

But all of our inquiries concerning these ancient inhabitants and their works are attended with the greatest difficulties. We are continually compelled to wander among great piles of earth presenting various tumuli, together with the effigies of men and animals
which were all covered with giant trees, whose only
voice is that of the western gale, as it whistles through
the dark and dismal forests, and beneath whose roots
lay in solemn silence, the crumbling bones of a departed people—a people over whose history, and through
whose works, hang the dark cloud of superstition; for
but few of their standing memorials (in the opinion of
archæologists) have been discovered but point to some
form of religious worship. But, as frequently observed,
the religious element is strong in human nature. It
is neither strange or wonderful that their greatest and

most magnificent mounds should in some way be devoted to the gratification of the highest passion implanted in the soul of man.

At first thought, it seems singular and wonderful that so great a labor should be supplied by a people (whose implements were mostly composed of stone) for the construction of such great works, addicted to war and superstition. But these ancient Mound Builders are not alone in building great and abiding structures dedicated to religious worship. The pyramids of Egypt, the most magnificent works ever constructed by human labor, and whose summits tower far above the highest monuments in Europe or America, have a religious significance. The great temple of Baal, located in the "capital of the valley," within the boundary of ancient Syria, was dedicated in the days of Paganism to the worship of Baal, or the sun. How far back, and at what time, the foundations of this great structure was laid is still a mystery; but it might have been contemporaneous with the Mound Builders, or perhaps more recent. All, however, is lost in the darkness of oblivion.

The ancient city of Babylon, noted for its former magnificence and grandeur, formerly contained colossal monuments dedicated to superstition and the worship of the gods. In the vicinity of the River Euphrates, near where once stood a bridge of gigantic proportions, was located the great temple of Belus. Herodotus, the father of history, describes it as it appeared in his own time, as occupying a space of two furlongs (about a fourth of a mile.) In the middle of the sacred precinct stood a solid tower, a furlong both in depth and width. Upon this tower stood another, and another upon this, to the number of eight towers, rising one above the other. In the last tower stood a golden table. During the night this sacred place was presided over by a native female chosen by the priest and sanctioned by the voice of the god whose presence was in the temple during the darkness of the night, reposing upon an easy couch. In one section of this great structure was a golden image designed to repre-sent the form and personal appearance of their fabled divinities. Near this image was a golden altar, and upon another altar of larger dimensions offerings of frankincense were annually made when they celebrated the festival of the gods.

In writing of any ancient people we find that war and mythical associations are the standpoint of their history. Every temple, every mound and structure, is but a silent chronicle of war, or some lifeless super-stition. America, in her infantile life, was afflicted with the curse. Baptists have been driven with fury from their homes to the primitive wilderness, and their property confiscated. The blood of innocent Quakers has been spilled, even in Boston. Men, women and children have been put to death for witchcraft—for a crime that had no existence.

ANAMALOUS MOUNDS.

Among the great amount of mounds and ancient

works so widely scattered, many present characteristics which we have been unable to classify.   Some present features that resemble temple mounds, but are
not well defined, many of which, no doubt, were used
for a double purpose.   On being explored, we frequently find a sepulcher and an altar, both representing the
general characteristics of the mounds that were devoted to the dead, and supposed to have been used for
sacrifice.

Also among the large number of anomalous mounds
which have been discovered may be included a number
located in Ohio, which some are disposed to call mounds
of observation. for the reason that they are always located upon the highest hills, from which can be seen a
great extent of country in either direction, and more
particularly, the valleys of rivers.   A range of these
peculiar and singular structures are found on the eastern borders of the Sciota valley, and extend from Chil-
cothe to Columbus, through the most fertile and picturesque section of the State of Ohio.   These works
are so situated and placed with regard to each other
that signals of fire could be transmitted the whole
length of the line in a few minutes.   Nearly opposite
the villuge of Chilicothe is a giant hill, the loftiest of
any in that section of the conntry—the summit being
elevated six hundred feet above the surrounding territory.   One of these mounds is placed directly upon the
top of this hill.   When the forests are stripped of their
foliage. as in the spring and autumn, a fire placed upon

the mound could be distinctly seen for the distance of twenty miles, both up and down the valley.

At Circleville, about twenty miles distant, commences a series of similar works, extending for some distance up two rapid streams. The highest elevations bordering these creeks contain those anomalous structures They frequently occur on the Ohio, Miami, Wabash, Illinois, and Mississippi Rivers, and in several places, such as Portsmouth and Marietta, mounds of stone have also been discovered, which were no doubt used for a similar purpose

At the distance of about four miles in a northerly direction from the village of Chilicothe is a singular work, which, to all appearance, was designed for defensive purposes, where the square and circle are closely associated. The figures are perfect, which would indicate that "the builders possessed a standard of measurement and had a means of determining angles." The object. of constructing this work in connection with the great chain of mounds is difficult to determine.

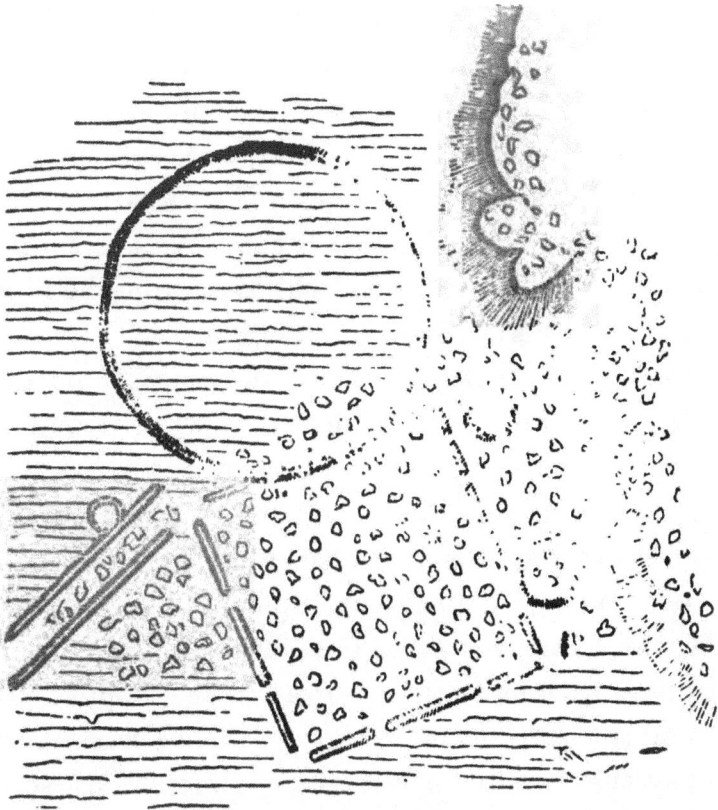

Fig. 20—Works north of Chilicothe combining the square and circle.

That these mounds located upon the highest hills and connecting the most important ancient towns or cities were mounds of observation, is the opinion of many. Marks of great and extensive fires are plainly visible. It is very likely that sentinels were kept around the mound continually ready to increase or light a fire on discovering a signal from either direction. In times of threatening danger, if such was the case, we can readily beleive that their means of communication were far in advance of us previous to the invention of the telegraph. It is not to be supposed that this singular mode of communication was used for common purposes, but for great and exciting occasions, such perhaps as the sudden approach of an enemy; then it would challenge the warriors to defend their country, their homes, their gods and their wives and children from the terrible snares set by a cruel and vindictive foe.

As this section of Ohio contains the most important and extensive works of the long lost people; and as several cities were located in the vicinity such as Circleville, which covers a large amount of territory, it is very probable that this novel and rapid mode of communication was used for other important business than that of war. Judging from the standpoint of American people previous to the telegraph, we would suppose it might have been used for political purposes and to convey the intelligence of the election of some prominent officer. These numerous people would naturally require some mode of communication other than by messengers sent through the different trails leading from city to city, and that they had some means connected with the mode of arranging their fires to convey different meanings is not impossible. One of the largest and most prominent mounds which is assigned to this class is located at Miamisburg and situated on a high hill but a short distance east of the great Miami River.

Fig. 21—The great mound at Miamisburg.

This mound must have been the grand centre of attraction when the first alarm was given, for far from its summit could be seen the great river valley and the ancient fort at Alexanderville, together with several others still up the river. This mound belongs to the largest class found in America. It is sixty-eight feet in perpendicular height, eight hundred and fifty-two feet in circumference at its base and contains nearly three million cubic feet of earth.

Admitting these mounds were used and constructed
for the purpose of conveying intelligence, which is the
most reasonable theory, it further goes to show beyond
a question that their builders were not the low and degraded people which many suppose.   These hill structures are very extensive and command a large amount
of territory and show great skill in their construction.
The soil of which many of them is composed, as is
generally the case, was brought from a distance, much
of it up steep and rugged bluffs, consequently they
must have been very expensive.   All the purposes to
which they were devoted we shall probably never
know, but as we wander among the ruins of this lost
and departed people, it is painful to behold the destruction that time and man has worked. Great changes have taken place since these imposing monuments
were sacred to a people who once outnumbered the
present inhabitants of that now densely populated
country.

The section where these anomaulous mounds are located must once have contained many cities.   Circleville was to them a place of great importance and no
doubt contained thousands of inhabitants.   It is located in a fertile and interesting section of the country
and seems well formed by nature for a large and central city.   When the white man first set foot upon the
soil it was a wilderness, but notwithstanding the trees
traces of former occupancy by man were plainly visible.
Streets extending in different directions were easily
traced, and the circles and streets were laid out with
geometrical percision.

The cities and villages once occupied by those ancient people were undoubtedly constructed of wood, the peculiar character of which will forever remain a mystery, for Time, that fell destroyer, has crumbled them to dust. The manner in which the foundations of their buildings were laid is difficult to determine, but it is very likely they were merely placed on the ground for we have no evidence that stone foundations were used. If such had been the practice they could be readily seen in the solitary wilderness. Some are inclined to the opinion that rude cabins were the only habitations of the Mound Builders, but with regard to that theory I am disposed to differ. All of the great works which have been discovered point to a people who were noted for skill and artistic design. The sepulchral mounds, such as the one at Moundsville, show remarkable care in their construction, for after standing unknown centuries they now show remarkable engineering skill. If those people were so particular with regard to their works which still remain, it is but reasonable to suppose they would be so with regard to their wooden habitations. Notwithstanding their implements were peculiar to the Stone Age, still they may have had some way unknown to us to build great and imposing structures. When the Spaniards first entered Mexico they were astonished at the wonderful works that presented themselves to view. Great piles of earth and stately buildings composed of wood were numerous and constructed with mechanical skill, and without the use of iron.

We have reason to suppose that the Mound Builders
were as far advanced in many respects as the ancient
Mexicans, and to all appearance Circleville might once
have been a town of as much importance as the city of
Mexico, the great capital of the Montezumas. If the
Mound Builders were the forerunners of the Mexican
Empire it is highly probable they would practice building in a style similar to the Mexicans, although some
of the great structures located in the capitol were composed of stone and lime. If we can believe the Spanish writers, the city of Mexico at the time of the conquest was hardly excelled by any other city for grandeur and magnificence. The palace of Montezuma was
a vast edifice with twenty doors to the public square
and streets. It contained three great courts, in one of
which was a beautiful fountain, together with several
halls and more than a hundred chambers. Some of
the apartments had walls of marble. The beams were
composed of cedar, cypress and other valuable woods,
carved and decorated in an artistic manner.

Of this great temple it is stated, "it stood in the centre of the city," and so great were its dimensions that
Cortez affirms that a town of five hundred houses
might have stood upon the ground which it occupied.
All the palaces of the king were beautified with gardens where was cultivated beautiful flowers, odoriferous
herbs and medicinal plants. Clavigero says: "Montezuma had woods enclosed with walls and furnished
with variety of game in which he frequently sported."

The palaces were all kept in the greatest order and

neatness, and notwithstanding that Montezuma was a pagan he took great pride in the cleanliness of his person ; hence he had baths in all of his palaces, and bathed with great regularity every day in the year.

Two large buildings in the city of Mexico Montezuma appropriated to animals ; one for birds of song and beautiful plumage and the other for birds of prey together with quadrupeds and reptiles. In the latter much space was devoted to strong and powerful cages where lions, wolves and various othes carnivora were confined.

From facts before us it appears that the Mexican people were in a high state of development when they fell by treachery of Fernand Cortez. They were skilled in various arts, such as the working of gold, sculpture, painting, etc.

The lords and leading officers of the government could vie with the lords of England in their palaces, gardens, woods and hunting grounds.

If the Mound Builders were driven from the north several thousand years ago to Mexico and became the powerful people which history claims them to have been, we can readily believe they once occupied cities in the north equal, if not superior to the great cities found in that empire.

If the inhabitants of Mexico should suddenly be destroyed, and after thousands of years had passed away another people possessed of our intelligence should explore their country it is doubtful if as many marks of their former greatness could be found as are now visi-

ble in the United States. Their works composed of
wood would soon crumble into dust, and those composed of stone and brick would present a mass of ruins.
It is very likely that the section where these anomalous
mounds are located large cities containing structures
made of wood were once extensive. The class of antiquarians who adopt the theory of rude cabins for the
Mound Builders form their opinion of the Indian races
who show no disposition to build anything grand or
magnificent. No tribe of Indians ever discovered
build large and elegant structures. The Mound Builders were a very different people who delighted in great
and imposing structures without regard to expense. If
the Mexicans could build great cities without the use
of iron it is very probable the Mound Builders could
do the same thing.

The theory that a race of people possessing the form
of head peculiar to these ancient people could merge
into the Indians, is not only untrue but unphilosophi-
cal. Nature shows progress in all her works, and has
written it everywhere.

CHAPTER IX.

SEPULCHRAL MOUNDS AND CEMETERIES IN TENESSEE
AND FURTHER SOUTH.

I have heretofore frequently referred to a class of tumuli many of which are of large dimensions, and distinguished as sepulchral mounds, which are found in all sections of the country formerly occupied by primative inhabitants. These works are very numerous and almost invariably have the form of a simple cone. They differ in their dimensions, being from six to near a hundred feet in height. They occur frequently within the walls of great enclosures, and are often found in groups in such close proximity as to be connected at the base. When they are found thus connected they vary much in size, being from six to forty feet high.

The object of constructing these mounds is much better understood than most of the works of the ancient people. They always contain a skeleton and frequently more than one, and were no doubt built to do honor to the memory of some distinguished character, who fell by that immutable law that surrounds all living beings who must soon follow that wonderful people through the "bitterness of death."

In connection with the remains found in these places of sepulcher, stores of various articles of great value always surround the crumbling skeletons; articles which were not designed to be hidden and then withdrawn, but to remain until the lifeless clay should put on the "garb of immortality."

Notwithstanding these sepuralchral mounds are so numerous, and can be counted by the thousands, it is not to be supposed that all of the great and extensive people were thus entombed. If various parts of this American Continent was as densely populated as archæologist believe it to have been, the earth could not have contained the vast number of mausolea which would have been required. The toiling millions through whose industry and labor the kings, priests, and warriors were sustained were allotted a more common burial, with naught but the "clods of the valley" to mark their humble tomb.

In many localities in the vicinity of their greatest works vast cemeteries have been discovered containing great numbers of decaying skeletons. These contained the remains of the great multitude of the plebeian people who died devoid of a "reputation and a name." Such cemeteries have been disclosed where the soil and chemical conditions were adopted to their preservation; many no doubt which were once extensive have now become obliterated. In several places on the River Ohio many ancient grave yards have been discovered. Mr. W. M. Clark in a paper describing the antiquities of Tenessee. says: "The most celebrated cemetery, and the one most frequently resorted to by relic hunters, is at 'Old Town,' seven miles northwest of Franklin on the farm of Mrs. Brown. Formerly, like other encampments, it had a wall and ditch surrounding it, but they are gone. There were many graves and mounds scattered over the enclosure. but most of these graves have long since been emptied of their contents, and the mounds for the most part have been dug into. However I obtained some very interesting relics here, among them two beautiful pieces of ivory carved with a precision seldom seen among Indians, they are made from the tusk of the mastodon. The larger one must have come from the tusk of a monster, for to furnish material for such a gorget, it must have been twelve inches in diameter. These gorgets have two holes in the edge, near each other, and they were most probably worn suspended on the breast, and may have been emblems of authority. A piece was fractured off one edge by accident after taking it up. A string of beads was procured here ; they are made of bone, are quite small and were lying in the grave of an infant. The dead in this cemetery were all buried with their heads towards the east, and some graves contained bones of three or four persons. It was quite common to find the bones of children and adults in one grave, though occasionally a grave was occupied by several children. The relics, when there were any, were always found by the side of the skull."

Among the great amount of relics entombed with the Mound Builders found in these ancient cemeteries, may be included stone axes in great numbers, and other implements peculiar to that people. In one of the graves opened by Mr. Clark were found five beads of amber of beautiful workmanship. They were about two inches long and near a half-inch in diameter in the centre. The material of which these beads are composed must have been brought from a great distance as it is found in alluvial soil at Cape Sable in the State of Maryland. The practice of wearing beads was not peculiar alone to the Mound Builders, but has been and is practiced by all nations and people of which we have any knowledge. Beads of a peculiar design and of valuable material have always been considered one of the greatest ornaments to charm and decorate the female sex.

The custom of burying articles of value with the dead has been practiced by nearly all the pagan nations. The Mexicans practiced it to such an extent that all of the graves of the rich contained gold and precious stones together with various other articles considered of great value. Among the relics which in many instances comprised their earthly wealth, they deposited various kinds of food which was designed to sustain them on their long and toilsome journey towards the shores of a physical immortality.

When the fact of the wealth contained in these tombs was made known to the Spanish conquerors they commenced to gratify their inordinate thirst for gold. The tombs of the lords and wealthy inhabitants were desecrated by that Christian people who claimed a de-

sire to instill into the minds and hearts of those degraded and wicked pagans the choicest blessings which
were to be found by the adoption of their holy religion. But gold, one of the greatest tempters, reigned
supreme, they became wild and insane for sudden
wealth, trampled under their feet the "golden rule,"
and desecrated the most sacred and magnificant tombs
to be found within the boundaries of the kingdom of
Montezuma. Cortez in one of his letters written after
his entrance into the Capitol, when it was beseiged by
his army says that one of his "soldiers found two hundred and forty ounces of gold in one sepulcher which
was in the tower of the temple."

The most interesting relics found in these ancient
cemeteries in Tennessee by Mr. Clark are several images wrought in sandstone and clay. One of these idols
weighs about twenty pounds. Among the four specimens found one only was composed of sand and clay
and that had been subjected to intense heat. The largest of these images says Mr. Clark, is remarkable for
its resemblance to the idols of India and China.

The workmanship is rude it is true, but faithful in
its details. The legs are only represented to the knees,
there is an attempt to show the hair, and at the back
of the head there is a knot of hair, with a loop for the
suspension of "ornaments." From the description of
the idols given by Mr. Clark, they have a striking resemblance to the image formerly mentioned found near
the mouth of Cold Spring Creek, on the Allegany River, some twenty-five years ago. The maker of that

image seemed to have exhausted his best skill to represent the hair hanging in ringlets. This image was not found in an ancient cemetery, but within a short distance from the Cold Spring mound which according to the testimony of those who explored it, contained cart loads of human bones in the last stages of decomposition.

The largest of these figures was designed by the sculptor to represent a man. This image was found lying in a grave by the side of a great skeleton much larger than the present race of men.

This ancient graveyard points directly to the Mound Builders, a people who undoubtedly were somewhat addicted to idolatary. Mr. Clark and other archæologists regard the images buried with the dead as idols. It is very likely the burying those images with the corpses of their people had some significance intimately connected with idolotry, but what it was is hidden in the darkness of the tomb.

The mode of ancient sepulchere is not entirely confined to mounds or graveyards such as has been formerly described.

CEMETERIES IN THE SOUTH.

In several of the Southern States the mode of urn burial has been practiced by some class of ancient people. In the group of mounds located on the banks of the Wateree River, in the vicinity of Camden, South Carolina, have been discovered vases with human bones

—where to all appearance the skeletons were packed after the flesh had been removed. Whether the corpses had been dissected and the flesh removed from the bones by artificial aid or by decomposition is unknown. Numerous cemeteries have been discovered where this kind of burial alone was practiced. One of the cemeteries devoted entirely to this mode of burial is located on the Island of St. Catherines on the coast of Georgia.

In the Northern States no evidence of urn burial has ever been discovered. Why it should be so extensively practiced at the South and not at the North is not understood. Perhaps it may point to a different empire or government, and also to the cause of building so many great military fortifications in the North.

It is not improbable, as we have formerly suggested, that there was once an ancient Northern and Southern confederacy. The wants of the people in the Southern States were different thousands of years ago from those of the north. The north held articles of great value to the south which could not be obtained without treading the northern soil. The great copper mines of Lake Superior, so valuable to that ancient people everywhere was reached through the north; for copper which was obtained in those mines has been distributed extensively in the south. And then the mica mines located in the south could not be reached by the northern people without infringing upon their soil. Copper and Mica were articles for which great labor was performed to obtain; and mica, as is so apparent, was regarded with religious veneration, as we nearly always find it surrounding the skeletons of humans.

Now the most reasonable theory is that the north and south were divided and those great fortifications in Ohio and in numerous places through the north were the result of long and continued war, which for aught we know might have been equal to that of the crusaders when they were fighting for centuries for the deliverance of the Holy Land.

The circumstance of the southern people pursuing a different mode of burial would indicate that their habits and customs differed, which would point to a different form of government. If those ancient people were joined in a single confederacy, it is reasonable to suppose they would adopt the same forms of religious worship, the same imaginary gods and the same mode of burying their dead. It is reasonable to suppose from the great number of sepulchral mounds so numerous in the Southern States that one form was once practiced throughout every section of country occupied by that strange and wonderful people. Why such a change should have taken place will perhaps always remain a mystery. But the theory that I have heretofore suggested that a portion of those people were driven to the territory bordering the northern Lakes appears very apparent. The military fortifications which have been discovered in several counties in Western New York can be counted by scores. They were built on a scale which would indicate that they were driven in different directions and fortified themselves as best

they could against the arrows of the enemy. Whether
the cause of this war that built so many fortifications
grew out of superstition (as many suppose) or some desire to wrest from the beseiged the copper mines of
the great lakes, or some valuable territory bordering
the great chain of inland seas, will forever remain a
mystery.

CHAPTER X.

MOUNDS, CEMETERIES AND OTHER ANCIENT WORKS LO-
CATED ON THE CUMBERLAND RIVER AND OTHER POR-
TIONS OF TENNESSEE.

The ancient cemeteries and earthworks located in
Tennessee are very numerous and can be found in almost every county. The central part of the State in
the vicinity of flowing streams and fertile lands discloses various works which were built long ages ago
by the ancient inhabitants. A remarkable chain extends from the Tennessee River diagonally across the
state to the Cumberland River, in Kentucky. These
works comprise cemeteries, mounds and military fortifications.

This section of Tennessee is remarkable for the richness of its soil, which repays the husbandman for his
toil, and bountifully supplies the inhabitants with the
choicest luxuries of life.

The sepulchral mounds which have been excavated
in this vicinity disclose human remains which show
evidence of great antiquity, and no doubt have slumbered in those great tumuli while the earth has performed more than a thousand journies around the central luminary.

Near the village of Franklin is a large sepulchral mound located upon a high hill which stands alone and from which can ke seen the circumjacent territory for many miles. The lands in the vicinity of this mound are the most fertile of any in the state, and in the season of harvest its broad acres are beautified with golden grain. This tumulus after being exposed for unknown centuries to the action of the elements (when opened in 1875) was twenty feet in height, four hundred feet in circumference and entirely surrounded by a terrace composed of soil. It was covered with trees of equal magnitude with those which surround it on the hill, the soil of which had never been disturbed by the white man. The hill upon which this mound is located, is remarkable for the great amount of limestone bowlders located upon its surface. When the mound was explored it was found to be composed mostly of stone peculiar to those by which it was surrounded. These stone were covered with a layer of earth near two feet in thickness. The stone which composed the remainder of the tumulus varied much in size, weighing respectively from twenty to more than a hundred pounds. About four feet from the top of this mound was discovered a layer of graves, which extended nearly through its whole diameter, where they were located. Says Mr. Clark, "the graves are of two wide slabs about two and a half feet long for sides, and with the bottom, head and foot stone of the same material, making when put together, a box or sarcophagus. Each of these coffins had bones in it, some of

women and children together, and others of men."—
The bones found in this sepulcher were very much decayed, and when removed immediately crumbled to
dust. The relics found with the bones consisted of a
string of beads composed of shell, which after being
exposed have the appearance of chalk, various pieces of
pottery were disclosed, but mostly in broken fragments.
The section of country surrounding this mound was
once densely inhabited by this strange and singular
people, who from all appearance were engaged to a
large extent in tilling the soil. Maize was undoubtedly extensively cultivated and formed a large part of
the food of those ancient Americans.

The sepulchral mounds as a general rule are located
in the valleys, especially those of large dimensions, although they are frequently found in the vicinity of
great military fortifications located upon hills. Such
however, are mostly of small dimensions, built, very
likely, to entomb some great hero, who fell while fighting for his people, his gods, and his love of country.
It is the opinion of archæologists that this section of
Tennessee, especially in the vicinity of this hill mound
and other groups which surround it at no inconsiderable distance, once contained large and prominent cities
and villages, which had been occupied as such for a
long period of time ; as many of the works located
there point far away into the midnight of the past.
The remains of the dead found in this rock mausoleum
placed upon the hill, were very likely those of some
distinguished family, who desired to lay their bones

upon the solitude mountain, so when they should awake to newness of life clothed with the same physical body, they could once again look far away upon the verdant plains and flowery fields of their native land.

At the distance of about two miles from the above-mentioned mound, and in full view, an interesting group of tumuli was formerly discovered, and though they have been much disturbed by modern cultivation some very interesting relics have recently been discovered.

About two miles to the west occurs another group of mounds, four of which are in line from north to south. The largest of these is twenty feet in height and was about twenty-four rods in circumference when explored. About five feet from the summit of this mound a layer of ashes was discovered, beneath which the earth showed evidence of having been exposed to intense heat. Charred wood was mingled with the ashes, which was covered with earth when yet on fire. Layers of ashes occurred every five feet until the original soil was reached. There were found two very singular relics which seem to differ from any heretofore discovered. One of these relics (which is composed of copper) was designed to represent the human face. The work was skilfully done, considering the rude tools with which they performed their mechanical labor This relic was composed of four distinct pieces beaten together with stone hammers the strokes of which are plainly visible. On either side is a separate piece firmly riveted to represent the ears. The nose, face and eyebrows are distinctly marked and well defined. It is sufficiently large to cover the face in both length and breadth.   The purpose for which it was made is another of the mysteries connected with that singular people.   It is the opinion of some that it was used to decorate an idol, and in some mysterious way was connected with their mode of worship, and that this burnished surface glittering in the sun would inspire the assembled multitude with pride and renewed veneration for the sublime and awful mysteries connected with their holy religion.

The circumstance of finding so great a quantity of relics devoted to religious worship, distributed in all sections of the country once populated by the ancient inhabitants, has led many to believe that it points to a people sunk in the lowest depths of degradation. But that is not necessarily the case. It is a fact established beyond controversy, as we have discovered in all ancient works, that man is a religious being.   All pagan nations have their different forms of worship and various forms and symbols are used in their religious devotions.

Relics and symbols are more or less used by the priests of every religion.   They are largely distributed among various Christian churches and have a religious significance.   One of the most powerful churches in the world, and one whose temples of worship excel all others in grandeur and magnificence, can display cart loads of trappings and relics which are brought to bear

in various religious ceremonies. This great and powerful church whose followers can be counted by scores of
millions, whose head claims to be one of the successors
to St. Peter, is located in the ancient capital of the Roman Empire. This Christian monarch, the great head
of this great church, and in its nineteenth century, can
always be seen with a cross, the sacred symbol of his
religion,—not made of copper like the relics of the
Mound Builders—but of the finest gold and interspersed with sparkling diamonds. A golden chain is at-
tached to this sacred symbol, which lies across his
breast pendant from his neck. The cabinet of this
successor to St. Peter is rich with the relics of saints
and angels, together with golden candlesticks, and also
images of our Saviour wrought in marble, representing
the life-blood streaming from his bleeding and cruel
wounds.

In one of the apartments can be seen the infant Re-
deemer reposing in the arms of his virgin mother, laying in the stable surrounded by oxen and other domestic animals ; also the crown of thorns which covered
the sacred head during that time on the cross.

These venerated relics are not confined to the Vatican, but are distributed through nearly all the great
churches in Europe and America.

Numerous churches in Europe display the *genuine*
crown of thorns, together with large portions of the
true cross on which the Saviour expired. Now if one
of the leading priests of the Mound Builders could
burst the bands of death and again walk upon the

shores of mortality, visit the Pope and the Vatican and examine the relics, he would surely be struck with wonder and astonishment, and would be as much puzzled to define their meaning as we are over the relics peculiar to his religion.

In writing these papers it is not my purpose to discuss the merits or demerits of any religion or of any religious relics or symbols. Suffice it to say that mankind will always have numerous forms of religious worship, which will be continued so long as man occupies the earth on which we live.

Another curiously constructed relic was found in the above-mentioned mound, which was also composed of copper, the material of which must have been brought from the far-distant shores of Lake Superior.

This singular relic had the appearance of an hour glass. It had two concave disks which were fastened together by a hollow stem and extended through its entire length. It was composed of native copper and made in such an artistic manner that no joints could be discovered where they were united. In connection with these relics was disclosed the bones of a man in such an advanced stage of decomposition that nothing remained but bone dust, with the exception of a portion of the jawbone, which evidently had been preserved by laying in contact with the copper, which formed an oxide from the body of the dead while in the act of decomposition. This relic had a string drawn directly through its center of sufficient length to encircle the neck and hang upon the breast. The cause of the remarkable preservation of this string is attributed to the influence of the copper, by which it was entirely surrounded.

The purpose for which this relic was constructed is very difficult to determine. From what remained of the superior maxillary bone it is evident that the person buried in the mound was very aged. The teeth were gone, and to all appearance the osseous alveoli was absorbed long before this aged Mound Builder fell by the hand of death. From the apparent veneration of these primitive people for their rulers, it is very probable that the person entombed with this relic was a priest. who had won by his piety and eloquence a prominent place in the hearts of his admiring people. It seems very apparent, as we have frequently observed. that those ancient inhabitants believed in the resurrection of the body, hence it is reasonable to suppose that the curious relic thus entombed was an emblem that pointed to some prominent position that he occupied with the people to which he belonged. The circumstance of finding copper relics in the great tombs in connection with the remains of persons, is invariably believed to point to some one of great distinction among their countrymen. Copper to the Mound Builders was like gold to civilized nations. In idea we can suppose that this remarkable person whose bones are a heap of ashes, has frequently stood before great congregations of his fellow-men, with this emblem of authority suspended on his breast, breathing forth. torrents of persuasive eloquence—scores of years before

the angel of death severed the brittle and slender cord
that bound him to earth ; and that the doctrine he
preached was the resurection of the body, and that after he had broken the prison doors of his earthly tomb,
he would meet his friends once again clothed with
flesh, and bearing upon his breast the sacred emblem
of his high and holy office which had been conferred
upon him by the gods whose thrones are in the distant
skies.

WORKS AT NASHVILLE AND THE VALLEY OF THE CUMBERLAND RIVER.

The location where the city of Nashville now stands
is supposed to have once been a place of great importance to the primitive inhabitants of Tennessee. Great
and extensive works are found in all parts of the circumjacent country. A writer for the Nashville *Times*,
speaking of the ancient and populous towns formerly
existing in that section, says : "It may not be generally known that the ground on which the city of Nashville now stands was once the site of an ancient and
populous town, yet such is supposed to be the fact.
Some of the most interesting antiquities of our State
are found along the lower course of Stone river but a
few miles from Nashville. A wide area of country
there is covered thickly with thousands of graves of
now forgotten people."

"But a few years ago several gentlemen made some
researches in the region and found their labors richly

rewarded by some striking discoveries. At a place known as Schell's Springs they found a mound of considerable height, and perhaps forty feet in diameter. which proved upon examination to be nothing less than a vast mausoleum of the dead. The graves were found to be made of flat rocks, symmetrically joined together, and three tiers deep from the base of the mound to its apex. Bones were found in a remarkable state of preservation, together with pottery and shells, found only in the Gulf of Mexico."

"A gentleman of acknowledged skill and authority in antiquarian matters, who conducted the researches a few days ago, estimated that the remains they unearthed could not have had a less age than six or seven hundred years. What is remarkable, no warlike implements are found in this locality, from which it is inferred that they were a peaceable race, and were probably exterminated or driven away by the Indians. In one of the graves was found a beautiful vase, that had been placed in the hands of the inmate of the tomb at the time of the burial. Upon this little bit of earthenware was the model of a diminutive animal. The care which these people took of their dead shows a high degree of humanity. We learn that other investigations in this section are to be made ere long. The field is certainly a rich one for the antiquarians."

The fact of finding such vast quantities of shells in many localities, and the use to which they were applied, is not generally understood among antiquarians. That they used them in a small way to make beads

and other peculiar articles to decorate their persons is
very apparent. The cart loads of shells found in the
vicinity of Nashville must have been very expensive to
these ancient people, especially when we take into consideration the great distance from which they were
transported. If they were brought by land the near-
est place at which they could have been obtained would
be Mobile Bay in the State of Alabama; hence they
would be required to travel the whole length of the
State, over rivers, hills and mountain summits, and
then on through a large part of Tennessee, ere they
could reach their home in the valley of the Cumberland. If they were transported by water they must
have been collected on the shores of the great gulf,
and thence up the winding Mississippi to its union
with the Ohio, and from there to its junction with the
Cumberland, a distance of more than one hundred
miles from the place of destination, and nearly fifteen
hundred from where the turbid waters of the mighty
river meets the waves of the ocean. The most reasonable theory with regard to the immense amount of
ocean shells found in such great quantities in the valley of the Cumberland, is that they were made an article of commerce by those ancient mound builders,
and that the present location of Nashville might have
been a great market for the sale of that, to them, valuable article.

From the vast number of those people who once occupied the great valley of the Mississippi together
with that of the Ohio and Cumberland, we can believe

that the time once was in the long ago when different tribes from neighboring cities might be seen wending their way over the Cumberland Mountains, or along the river valleys, driving the huge mastodon, (or some other animals now extinct), loaded with the riches of the ocean and making a tour of traffic similar to the ancient Arabians while traversing with loaded camels the arid and scorching sands of the African desert.

About two miles from the city of Nashville, on an eminence which commands a view of the Cumberland river and also a large extent of the adjacent country, is located a vast cemetery rich with the ancient remains. The ground in the vicinity of these graves, denuded by the action of the elements, is literally covered with marine shells. The graves where lay the crumbling bones are composed of stones which were deposited around the dead in which they are coffined.

One of the relics found in a grave excavated by Mr. Robertson was that of an earthen vase with perforations on either side so as to be suspended with a string or chain from the neck. This grave contained two skeletons, one of an adult and the other of an infant. The bodies, when laid in that tomb of stone, were covered with ocean shells which have undoubtedly had much to do with their remarkable preservation. The vase was found near the left shoulder of the larger skeleton, and was filled with some carbonized substance, the nature of which has not been determined ; but very likely it was designed to be used for some

purpose unknown to us, and used once more when the
body purified by the sleep of death should again roam
over the hills and through the enchanting valleys,
meeting friends and kindred who long had been their
companions in the silent tomb.

These graves are very numerous in all directions
immediately surrounding Nashville, and also within
the limits of the city.   On an eminence near a remarkable mineral spring, scores of graves can be found and
also large quantities of pottery which is scattered in
all directions.   From the general appearance the vessels must have been very large and are supposed to
have been used for the purpose of evaporating salt
from saline springs located there.   The circumstances
of finding salt springs in that locality is undoubtedly
one cause of their once extensive population in the fertile valley of the Cumberland.   Those unknown people seem to have possessed a remarkable skill in discovering various articles upon which they placed great
value.

THE WONDERFUL ABILITY OF THE ANCIENT PEOPLE IN
SEARCHING FOR HIDDEN TREASURES IN VARIOUS
SECTIONS.

When the great copper mines of Lake Superior were
explored by S. O. Knapp, the agent of the mining
company, he was astonished at the wonderful ability
displayed by an unknown people in searching out
great veins of native copper, which were hidden far

below the surface rock. From all appearance it would seem they were guided by an instinct which led them to the richest fields ever found in that most fertile copper region yet discovered. Perhaps they were spiritual mediums, whose exalted visions could discern the valuable treasures deeply hidden in the bowels of the earth. If such was the case they were surely crowned with much greater success than the mediums of the present day. The great size of the vessels which have been disclosed and the material of which they were made, would show that their mode of producing salt from the saline waters of the spring, was by solar evaporation, for it never could have been done by the present mode of evaporating the water by artificial heat. Pottery of various kinds is found in nearly all sections of the country once occupied by these lost and unknown people, but not as a general rule of so large dimensions as are found in the vicinity of saline springs. They displayed remarkable skill in the manufacture of earthen vessels, for they can now be found in broken fragments in plowed fields, which have been cultivated for a hundred years or more in a good state of preservation.

So extensive are the ancient works, and presenting such a great variety, we are left in the dark in forming an opinion of the use to which many were adapted, and also of the manner in which their numerous discoveries were made.

DISCOVERY OF A SUBTERRANEAN FOUNTAIN IN IOWA.

In Wright County, State of Iowa, is a walled Lake, located about one hundred and fifty miles from Dubuque City, which is thus refered to in an editorial in the *Dubuque, Iowa, Herald.* "The greatest in the State of Iowa, and perhaps any other State, is what is called the Walled Lake, in Wright County, twelve miles north of Dubuque and Pacific Railroad."

The lake is from two to three feet higher than the earth's surface. In some places the wall is ten feet high, fifteen feet wide at the bottom and five feet at the top. Another fact is the size of the stones used in construction, the whole of them varying in weight from three tons down to one hundred pounds. There is abundance of stones in Wright county, but surrounding the lake to the extent of five or ten miles there is none. No one can form an idea as to the means employed to bring them to the spot or who constructed it. Around the entire lake there is a belt of woodland half a mile in length composed of oak, with this exception the country is a rolling pararie. The trees must have been planted there at the time of building the wall. In the spring of the year 1856, there was a great storm, and the ice in the lake broke the walls in several places, and the farmers in the vicinity were obliged to repair the damages to prevent inundation. The lake occupies a ground surface of 2800 acres; depth of water as great as thirty-five feet. The water

is clear and cool, soil sandy and loamy. It is singular
that no one has been able to ascertain where the water
comes or where it goes, yet it is always clear and
fresh."

For what purpose the vast amount of labor in constructing such a great wall was performed is surely
one of the mysteries connected with this ancient people. The amount of stone which was used would comprise thousands of cords; and the fact of their heft and
the distance from which they were transported, would
show conclusively, that they must have been aided by
some powerful beasts of burden. Three tons would
be too great a load for the most powerful horse teams
in this country, unless they had a plank or macadam-
ized road, in the best of order. But the building of
the wall was not the most wonderful for if the lake
was then in the same condition as at present, without
a wall the surrounding lands must have been completely flooded with water. It is the opinion of some that
they possed a kind of instinctive knowledgle by which
they discovered mines of copper, mica, and the numerous saline springs from which they obtained their
salt; and by the aid of this remarkable faculty they
detected the location of a hidden fountain, built their
wall, dug down to the latent waters which were
driven up with great force and made the lake. This
theory possess too much of the marvelous for us, for
we are led to believe that the present inhabitants have
as much intuitive knowledge as any which have here-
fore preceded them.

CHAPTER XI.

SYMBOLICAL MOUNDS EMBRACING A LARGE AMOUNT OF TERRITORY IN WISCONSIN.

In the lower counties of Wisconsin extending over a territory about one hundred and fifty miles in length, by fifty broad, occurs hundreds of symbolical structures. They represent man, the lizard, bird, frog, fish, cross, crescent, angle, straight line, tobacco pipe, war-club and various animals and things which it is difficult to characterize. These singular effigies are located in the most fertile and interesting sections of the great State. They are never found isolated, but in groups which appear to have been placed by design with regard to each other.

On the borders of a small lake, which the Indian tribes regarded with religious veneration and named it the "lake of the great spirit," occurs numerous effigies, and among them is one that represents a bird of monstrous size with expanded wings. The body is one hundred feet in length by thirty in breadth; the wings from where they join the body measure respectively three hundred feet. The head of this wonderful effigy is well defined with a beak fifteen feet in length. In the immediate vicinity of this large work

occur various others of smaller dimensions, which somewhat resemble the eagle of the present time.   Near
the wings of these birds occurs an   effigy that bears
some resemblance to a deer, and on the opposite side
is another which is supposed to represens the bear.

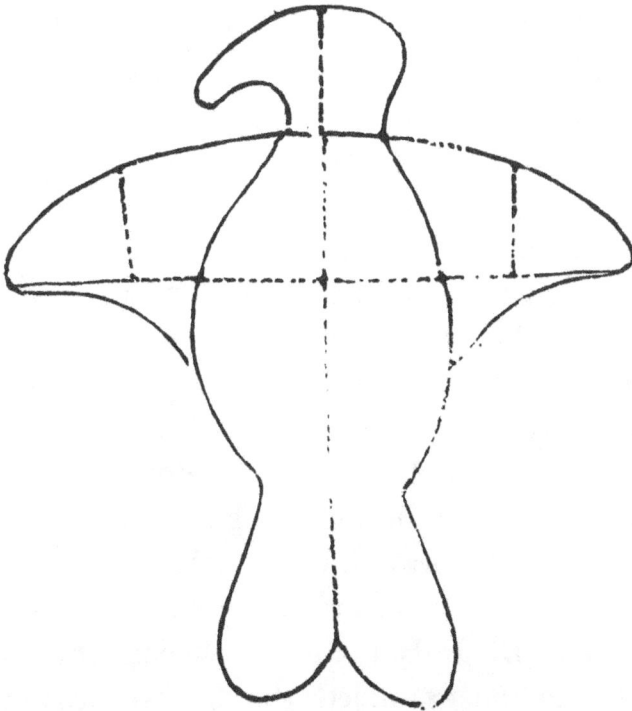

Fig. 22—Bird-shaped mound composed of eai th.

Sepulchral mounds are numerous in the vicinity of
these strange symbolical structures and present a singular mode of burial not usually found in other localities.

Near the beautiful lake Mendota, which is noted for its great variety of fish, is located a number of tumuli, which have the appearance of great antiquity. The largest of these mounds stands upon an eminence where it commands a view of the surrounding terri- tory for many miles. When this mound was explored a shaft was sunk from the apex down to the original soil. At the depth of about five feet occurred a layer of stones, below which was a bed of clay some four feet in thickness ; still below the clay another layer of stones similar to the first was disclosed. These were formerly taken from a stratum of limestone below the shore of the lake, and showed indications of the action of the waves which dashed against them long ages ago. Below these stones was found a bed of ashes mingled with charcoal, together with flint implements and various shells. About a foot below this layer of stones was discovered a human skeleton which to all appearance had been buried in a sitting posture. The bones were much decayed, though the inferior maxil- lary, several of the vertebra and the right humerus were entire. Dr. DeHart, the gentleman who explor- ed the mound and examined the bones anatomically, says : "The bones show great antiquity, and the ver- tebra were larger than those of the present type." He further says the humerus was perforated at its inferior extremity.

In different states, especially where these effigies are located, it is not uncommon to find the mouldering bones of men perforated.

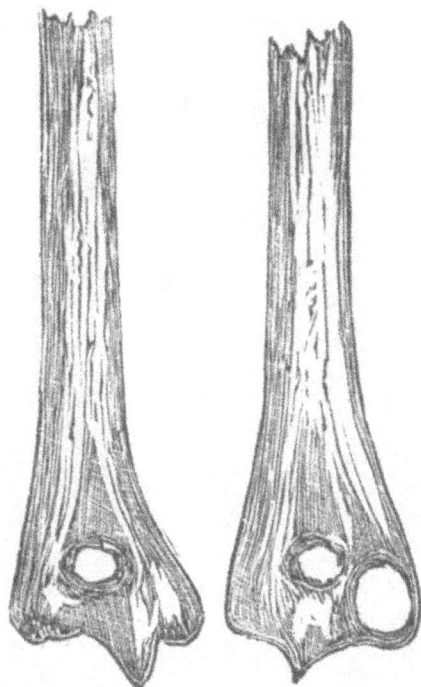

Fig. 23—Bones exhumed by Dr. DeHart.

In many mounds (especially several in Michigan) the long bones have been found perforated in various places. So plain does it appear that these perforations, which extend entirely through the bones, were not caused by decay that the action of the drill, which was composed of copper or flint, can be readily seen.

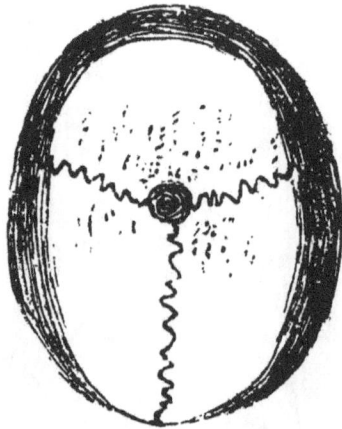

Fig. 24—Perforated Cranium found in a Mound.

The object of this singular performance, which must have been done upon the fleshless bones of the dead, is one of the numerous mysteries that present themselves to the archæologist.   It is, however, the general opinion that it points to some superstitious rite connected with the sepulture of the dead.   It might have been performed upon the living, by tearing the flesh from the bones while the sufferer was in the agonies of death.

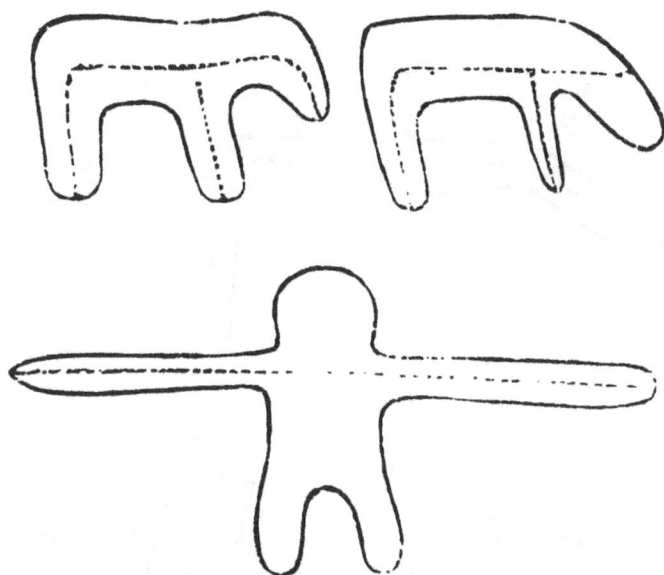

Fig. 25—Animal and Man mounds in close proximity.

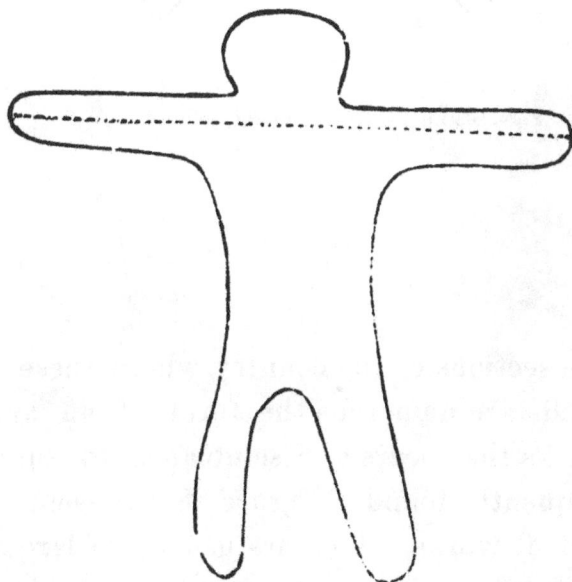

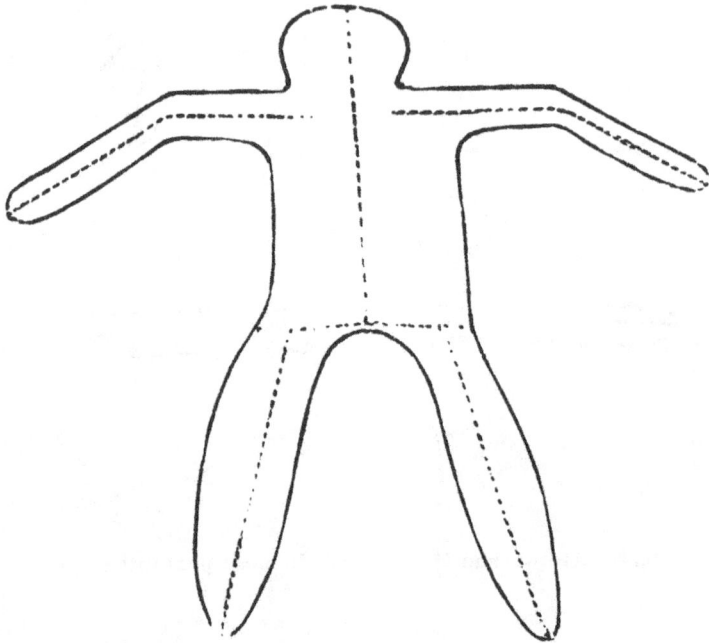

Figs. 26 and 27—Man mounds isolated.

In various sections of the country where these symbolical mounds are numerous the form of an animal
devoid of tusks that bears a resemblance to the elephant is frequently found.   Figure 28 represents one
of these ancient works, which are usually of large dimensions.

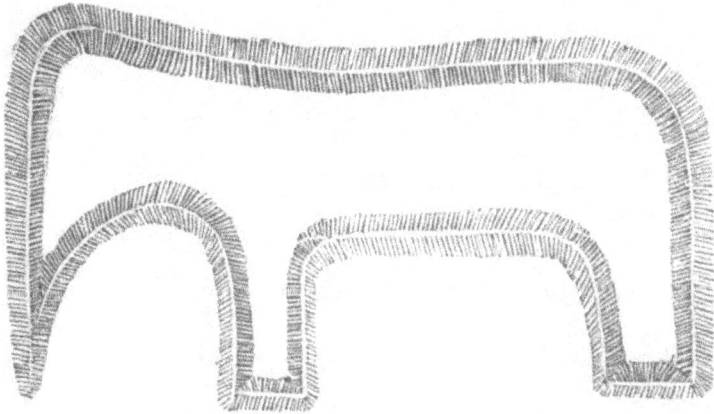

Fig. 28—Elephant mound.

The great elephant that inhabited this country after the close of the glacial period was clothed with wooly hair which was a protection from the severe cold. He not only inhabited the American continent but roamed amid the semi-artic woods and wilds of Europe, the companion of ancient man, and their bones are now crumbling together in the ground.

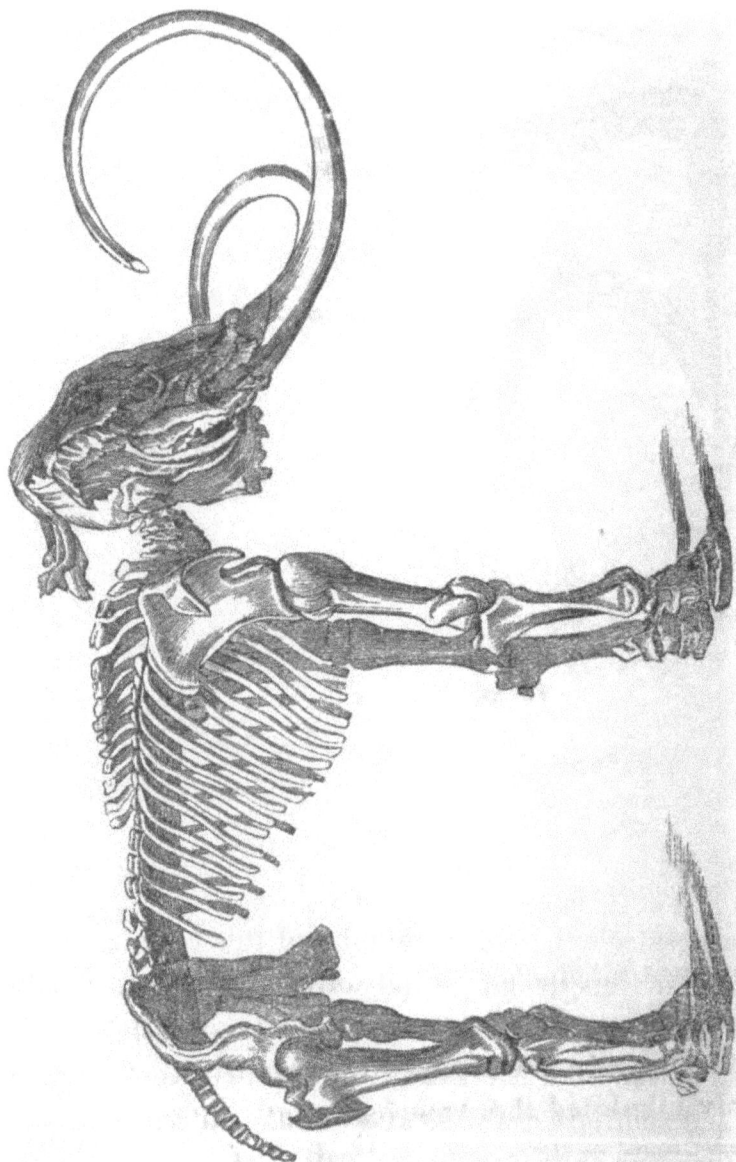

Fig. 29—Restored view of the Skeleton of the Ancient Elephant or Mammoth.

Throughout this fairest portion of Wisconsin, where the effigies are so numerous, the country is interspersed with flowing streams, copious springs and crystal lakes, and at one time to all appearance was the place where abode a happy and numerous people.

Various mounds are scattered in the vicinity of the lakes and rivers, which cover the bones of the dead; but the effigies so widely distributed, and built with so much labor, was no doubt pleasing to their superstition. It appears very probable that those ancient people were divided into clans, and each adopted different forms of worship. But if these singular effigies were symbolical of their religion and superstitions, as many suppose, it would seem to point to a low and degraded people. But religion not only among the most degraded pagan nations, but among the higher and more elevated classes, has adopted many fantastic forms of worship. The religion of the Arabian prophet, whose followers outnumber the mighty hosts connected with nearly every other religion, have built many strange and singular structures, which to a people who should come after them, ignorant of their history, would be puzzling to interpret; and then again, their religious faith would cause a scornful smile upon any intelligent countenance.

The Mahomadans adopt, with intemperate zeal, many strange and unnatural dogmas that they accept with pious veneration and unwavering faith; and also build singular and imposing structures to flatter their superstitions.

The followers of Mahomet are a powerful and intellectual people. They write books, build railroads and
do a large commercial business with the different nations of the earth; they build great monuments devoted to their religion and superstitions, which will stand
for ages after the frosts of time have converted their
common dwellings to a heap of ruins.

The section of Wisconsin where such a great amount
of ancient works are located (such as symbolical and
sepulchral mounds) was once, no doubt, highly cultivated by the ancient people, and as we find no evidence
that the land was cultivated in the vicinity of Lake
Superior, it is very evident that the products were
transported through the dark and dreary wilderness to
the border of the great lake to supply with food the
army of miners who were searching for the invaluable
treasure. The manner of transportion from the rich
and fertile lands of the Mississippi river and the garden lands of southern Wisconsin, to the copper mines,
is difficult to determine, but that the Wisconsin, St.
Croix, Ontongon and Mississippi Rivers formed some
of the highways, is probable. Other routes might have
been by land where is located numerous lakes and
ponds which would supply them fish and in part serve
them with food on their long and toilsome journey to
their copper Eldorado. The great wilderness country
bordering Lake Superior on the south for a hundred
miles or more is noted for small lakes which form the
head waters of numerous rivers which mingle their
waters with those of Lake Superior.

The great number of effigies located in the southern part of Wisconsin had prominent meanings which very likely caused the great diversity of symbolical structures peculiar to the different tribes from the north, the south, the east and the west, who were on their way to the copper regions in search of the precious metal. The section where the effigies are located might have been a place where they paid tribute by virtue of some treaty for the privilege of working the Northern mines, as those copper mines of Lake Superior are the only ones which have been discovered that show the work of these ancient people, and, as the copper was regarded of so much value we can imagine that the circumjacent country was swarming with pilgrims on their way to the shores of the great inland sea.

But in Wisconsin, and on the shores of the great lakes, the wigwams and rude habitation is gone. Another people now hold empire over the vast territory where once the dark and dismal clouds of ignorance and superstition skirted the mental horizon. The rude habitation of the savage has given place to the princely mansion and stately dome. Civilization has stepped amid the towering forest, and towns and cities have risen upon the dust of ancient ruins. The great Mississippi River, on whose waters once darted the bark canoe, now yields obedience to the steamboat that bids defiance to its rushing waters. The great lakes, once so sacred to the ancient people, and navigated only with rude vessels, now display the snow-white canvas fluttering in the breeze above the proud keel of the

merchant-man.   Locomotive engines, hauling their
heavy freight, now dash through the praries with
lightning speed, and then on, on, over great rivers and
beneath the mountain summit.   If the tombs of those
ancient people could give up their dead and place them
upon the shores of mortality, what wonders would be
presented to their astonished vision.   Man is a progressive being, and the simple word progress tells of
the wonders of this nineteenth century.

CHAPTER XII.

ANCIENT WORKS IN NEW AND CENTRAL MEXICO.

Vestiges of an ancient people and prominent marks of their civilization are disclosed not only from the northern lakes to the Gulf of Mexico, but are found to extend through every inhabitable section of modern Mexico, Gautemala, Yucatan, Honduras, and in nearly every part of Central America. The race who held empire over that vast domain at the time of the conquest have been denominated Aztecs. They were the humble and willing subjects of the great and powerful Montezuma, whose life and glittering glory went out amidst the carnage produced by the Spanish conquerors. The Aztecs were not the only race of ancient people that have lived, flourised, and occupied the territory where now stands Mexico. Another race preceded them, to which the name of Toltec has been given. They have left monuments far more imposing, and which presents a higher civilization than any other people who have come after them.

In the southern part of Mexico great and melancholy ruins of majestic temples and of large cities attest the former existence of a numerous and cultivated people ; but who they were and whither they went is as much of a mystery to us as that which surrounds the Mound

Builders, who have left such numerous and prominent works in the great and fertile valleys of the United States.

More than three hundred years ago, when the Spanish armies first invaded Mexico, scores of ancient and abandoned structures were discovered which seemed to indicate a higher state of development than those occupied by the Aztecs, but who the builders were was unknown to the Mexican people and was without a reliable tradition.

In northern Mexico the ancient works are not so extensive as in more southern portions, but in various sections solitary ruins can be seen at the present day. West of the Rio Grande and near the head waters of the San Jose and Zuni rivers, are the remains of numerous and extensive ancient works which show evidence of great antiquity. These works are built of stone, and so arranged in the wall as to convince the most skeptical of their ability in building great and prominent structures.

On the summit of a high bluff which is elevated some two hundred feet above the surrounding country stands the ruins of a great edifice built of stone. The foundation is nearly perfect at the present time, notwithstanding the vast number of centuries which have rolled into eternity since the superstructure was laid. The stones are uniform in size, being fourteen inches long and six in width; they are placed in horizontal layers in such a manner as to mechanically break joints with each successive layer below. In parts of this ancient structure (where not exposed to the action of the elements) are the remains of ceder beams which are well defined, though they show evidence of the destroying effects of time. Obsidian arrow heads, and pottery designed in artistic style, together with various other relics have been found mingled with the ruins.

In the beautiful and picturesque valley of Ojo Pescado are the remains of two great and ancient buildings, of such remarkable antiquity as to be far beyond the reach of the traditions of the present races of men.

These works are located in the vicinity of a copious spring, which shows artificial stone work in a good state of preservation. They are respectively about eight hundred feet in circumference and composed of stone, but the frost of more than a thousand years has crumbled them to heaps of ruins. The pottery found in the vicinity of the ruins show a high state of developement and fanciful taste; much of it was highly polished and decorated with bright colors in checks, bands and wavy stripes, together with the effigies of frogs, butterflies, and other harmless animals.

In the year 1854, Lieutenant Whipple while exploring in New Mexico for a railroad route to the Pacific, gave a description of the ruins of an ancient town at Zuni, which is now inhabited by Indians. This town was a ruin long ages ago, but in many places the Pueblo Indians built upon those crumbling and desolate walls, and now occupy the territory once controlled by a race of men much higher in the intellectual scale and

in civilization, than the present occupants. This ruined town, located amidst the most enchanting scenery, and in the fairest portion of earth, presents a
melancholly grandeur. Its crumbling walls soiled and
worn by unknown ages of time, lay in great confusion
over many acres of ground. The primitive foundations, notwithstanding the vast period which has passed away since they were laid, are firm as the hills and
attest the mechanical skill and ability of those who
laid them. The ancient walls and foundations which
now remain, are about six feet in thickness, and laid
in cement which seems as firms and enduring as the
rock by which it is surrounded.

The builders of this Zuni town, and many others in
Mexico and the surrounding country, were undoubtedly the Toltecs whose character and history seems
shrouded in darkness, notwithstanding Cabrera, Tor-
guemada and other Spanish writers have had much to
say about the Toltec migration, and about Huehue
Tlapalan which they claimed was the country from
which they originated. Their history is not only lost
to us, but was lost to the Montezumas centuries before
the fall of their magnificent and glorious empire. This
ancient town or city has been rebuilt and deserted several times since the primitive foundations were laid.
Those people who rebuilt the town reduced the walls
where broken down to the thickness of about twelve
inches. Small blocks of sandstone were used by the
latter builders which were dug from among the great
heaps of debris widely scattered among the ruins.

Among the great amount of relics found here may be included beautiful specimens of pottery, and elegant ornaments made of shells, indigenous to the ocean. Sections of ceder beams and posts elegantly carved, which showed evidence of great antiquity were found among the ruins. The location of this town is now sacred to the present Indian inhabitants, and consecrated to their religious worship. When Lieutenant Whipple was about to leave, his guide (one of the most intelligent of the natives) turned his face and "blew a white powder towards the altar three times, muttering a prayer, which he claimed was asking a blessing of Montezuma and the sun." It would seem from this performance that the religion of the Aztecs was ingrafted upon the present race of Pueblo Indians, who now occupy the country once so glorious among the ancient Pagan nations; but it seems improbable that the present natives can be genuine decendants of the Aztects, whose delight was in magnificent structures, elegant gardens, crystal fountains, flowery fields, trailing vines, and the luxuries peculiar to the highest and most cultivated people.

In the vicinity of the Colorado Chiquito, stands a great and magnificent structure now in ruins. The walls are three hundred and sixty feet in length, by one hundred and twenty in width, and in many places are ten feet in thickness, with rooms inserted in them. Relics in great abundance are found among the scattered fragments, and crumbled walls. Stone axes and decorated pottery is disclosed in large quantities. Indented pottery of various patterns has also been found, which shows the highest of mechanical skill displayed in its construction; and for artistict design and elegant workmanship, will challenge the best efforts of the nineteeth century.   This work is very old, and a ruin (no doubt) when the Aztec kingdom was in its glory, and long before it was exposed to the cannon and gunpowder of the Spanish army.

ANCIENT STRUCTURES IN THE VALE OF MEXICO.

As we go into Central Mexico we are there presented with wonderful ruins which attest to the great and numerous population by which it was once inhabited. In the northern part of the valley formerly stood the great city of Tulha, the "ancient capital of the Toltecs." This city was in ruins at the time of the conquest; by whom and when it was built was a mystery to the Mexican people more than three hundred years ago.

A remarkable pyramid located on an eminence which overlooks the fertile and beautiful valley, bordering the ancient town Xochicalco, presents a grand and picturesque appearance. The hill upon which it stands consists of rock "which was excavated and hollowed for the purpose of galleries and chambers." Five of these chambers are constructed within the pyramid, which are about six feet in height. Each of these galleries is covered with cement which was prepared from some very durable material, and so mechanically distributed that it now (after unknown ages) presents a smooth and glistening surface. The main gallery of this structure is one hundred and eighty feet in length, and terminates at two great pillars which were hewed from portions of the rock. Over the inner chamber stands a cupola about six feet in height which was constructed of hewn stones and laid in cement.

The vale of Mexico abounds in great and magnificent ruins. In this section stands the remains of the wonderful pyramid or mound of Cholulu, which, when measured by Baron Humbolt, eighty years ago, was fourteen hundred feet square at the base, and a hundred and sixty feet high, and covered an area of forty-five acres. This great work was terraced on either side, and constructed with four stages. But it is now a ruin, and the holy temple once so sacred to a lost and forgotten people has become the home of wild beasts, and poison reptiles inhabit the sanctuary of the gods. The surrounding territory, once the home of thousands, is marked with desolation and ruin. The antiquarians, curious to learn the history of this ancient city and its inhabitants, gaze upon the wonderful solitude; but the voice of those who built it is hushed in endless silence. The only voice is the broken and scattered fragments of happy homes, mingled with the ashes of the dead.

CHAPTER XIII.

MAGNIFICENT RUINS BURIED IN THE GREAT SOUTHERN

WILDERNESS IN CENTRAL AMERICA.

In Central America and the southern part of Mexico the ancient works which have been abandoned and left to decay are very numerous, and were marked with ruin long before the Aztec Kingdom was established. Many ancient cities are found buried in the depths of the forest, which were hidden from observation at the time of the fall of the empire.

The ruins of Palenque, which are very remarkable, were unknown to the Spaniards and the Aztecs at the time of the conquest. A great forest covers a large part of Yucatan and stretches into Gautemala, Tabasco and Honduras. Trees, underbrush and trailing vines peculiar to tropical countries impede the progress of the explorer at every step. So difficult to penetrate is this southern wilderness that many parts of it have hardly been trodden by the feet of man since the ground which it occupies was long ago cultivated by a people who have left such wonderful traces of a remarkable civilization.

In the southern part of this dense and tangled wilderness is located the ruins of Palenque. One of its most prominent buildings stands on a terraced foundation which was built in the form of a pyramid. The superstructure is forty feet in height, three hundred and ten feet in length and two hundred and sixty in width. The edifice which stands upon this foundation is two hundred and twenty-eight feet in length, by one hundred and eighty in width and twenty-five in height. It has fifty doorways, fourteen on each side and eleven at each end. This great and elegant building, whose foundation is forty feet above the surrounding plain, is built entirely of stone and lime. The great stones of which it is commposed are hewn in an artistic manner, and laid with great and almirable percision in mortar composed of sand and lime, which is as hard as the rock itself. An arched corridor nine feet in width surrounds the building externally on either side. Four courts of beautiful and elegant workmanship were disclosed in the interior of this ancient and wonderful palace. The largest of these courts is seventy by eighty feet in extent. All of these are surrounded by corridors which for architectural beauty and design stand to-day among the proud memorials of human genius.

The great mound on which this palace stands is cased with hewn stone which presents the most superior workmanship. Mr. Stephens says : "The piers around the courts are covered with figures in stucco or plaster, which where broken reveals six or more coats or layers, each revealing traces of painting." This circumstance would indicate that the building was con-

structed long before it was deserted, for it is very likely these renewals of plaster and paint were made at
far different periods.   The point of the greatest interest to the antiquarian connected with this deserted
palace, is the beauty and grandeur of its external ornaments.   The embelishments show the highest degree
of skill in architecture.   Human figures, single and in
groups, together with various fantastic ornaments engraved upon the polished rock, add beauty and interest to the magnificent structure.

Fig. 30—Palace at Palenque

Another building which has long withstood the waste
of time—although smaller in size than the palace—for
ornamentation of its walls, cornices and piers presents
to the beholder the most wonderful artistic skill.   Mr.
Stephens, referring to the human figures, says: "In
justness of proportion and symetry they approached
the Greek models."   This building is also constructed
upon a truncated mound or pyramid.   It is two stor-

ies in height and beautifully ornamented with stucco work and also inscriptions of tablets engraved upon the walls.

Fig. 31—Elegant structure at Palenque.

The most prominent and attractive object which presents itself within this structure is several human figures in bas releif which stand near a cross which is beautifully sculptured. This cross, to all appearances, was designed as the centre of attraction to those surrounding it, for a child is represented as being held towards this decorated emblem by one of the figures within the group.

The finding of this sacred symbol of Christianity in

this great American wilderness, in a magnificent temple decorated and embelished in a style unknown to
the inhabitants of Europe, has caused much discussion
among prominent antiquarians throughout the world.
Mr. John D. Baldwin, in his popular and interesting
work, "Ancient America," says : "The cross is one of
the most common emblems present in all the ruins.
This led the Catholic missionaries to assume that a
knowledge of Christianity had been brought to that
part of America long before their arrival, and they
adopted the belief that the gospel was preached there
by St. Thomas.   This furnished excellent material for
the hagiologist of that age, but like everything else
peculiar to these monkish romancers it betrayed great
lack of knowledge.   The cross, even the so-called Latin
cross, is not exclusively a Christian emblem.   It was
used in the Oriental world many centuries (perhaps
milleniums) before the Christian era.   It was a religious emblem of the Phœnicians, associated with As-
tarte, who is usually figured bearing what is called a
Latin cross.   She is seen so figured on Phœnician coin.
The cross is found in the ruins of Nineveh.   Mr. Layard, describing one of the finest specimens of Assyrian
sculpture, the figure of 'an early Nimrod king,' (he calls
it) says : 'Round his neck are hung the four sacred
signs, the cresent, the star or sun, the trident and the
cross.' These signs, the cross included, appear suspended from the necks or collars of Oriental prisoners
figured on Egyptian monuments known to be fifteen
hundred years older than the Christian era.   The cross

was a common emblem in ancient Egypt and the Latin form of it was used in the religious mysteries of that country in common with a monogram of the moon. It was to degrade this religious emblem of the Phœnicians that Alexander ordered the execution of two thousand principal citizens of Tyre by crucifixion. The cross as an emblem is very common anong the antiquities of Western Europe, where archæological investigation has sometimes been embarassed and confused by the assumption that any old monument bearing the figure of the cross cannot be as old as Christianity."

Palenque, from all appearances, must once have been a great and populous city. A writer who spent much time in making explorations says: "For five days did I wander up and down among these crumbling monuments of a city which I hazard little in saying must have been one of the largest ever seen."

It is very difficult to arrive at any correct conclusion in regard to the size of those ancient cities. Those which present the most grand, capacious and enduring structures are all buried in that sublime and solitary wilderness. The forest where the ruins of Palenque are located covers as much territory as either of the states of Ohio or Pennsylvania, and have, no doubt occupied that position for more than two thousand years. The common buildings occupied by the great majority of the inhabitants of those ruined cities were made of wood or some other material which has long since perished, the remains of which have left no more

trace of the place they occupied than the leaves of the
the forest that surrounded them a thousand years ago.

In surveying this wonderful city we are continually
presented with relics which display the most remarkable skill in their construction. Palenque once contained an aqueduct for carrying water to the city.   The
floors of the palace are composed of cement which was
laid in a mechanical and artistic style and for dura-
bility will compare with the foundation rock.   In the
excavations walks have been discovered composed of
hewn stone laid in mortar.   Vessels of   earthenware,
surprisingly beautiful, are   strewn upon   the ground,
covered only with the dust of ancient forests.   The
massive walls of great and ruined buildings, surrounded with trailing vines and overhung with the branches
of huge trees, lay scattered on the ground where once
stood Palenque.

It is mournful to reflect that ages ago this almost
impenetrable wilderness was active with human beings
zealously employed in cultivating the soil to sustain
the great population by which it was inhabited.   The
streets of the great city were alive with a busy throng ;
friends saluting each other with looks of love ; children sporting around the terraced foundations of those
enduring structures ; statesmen assembling in the
great hall of the palace discussing the best means to
perpetuate their glorious and magnificent empire ;
teachers instructing the youth in the mysteries of their
symbolical language ; old men and women tottering on
the verge of the eternal world wending their way to

some holy temple devoted to religious worship ; thousands of artisans hewing the rock for the construction of temples and other imposing structures; tillers of the soil distributing the golden fruits and the various products of the fields; merchants and artisans displaying their goods to the gaze of the busy throng that fill numerous streets; soldiers stationed at some favorable location which would command a view of the circumjacent country and detect the approach of an invading enemy; mourners carrying the lifeless bodies of friends and kindred to some consecrated tomb.

But now how changed. The glory of the mighty city has departed. Time, that fell destroyer, has laid her in ruins. Where once was a living multitude, joyous and happy, now stands the solitary wilderness, with no tongue to tell its history. The silence of death is substituted for the noise and bustle of the thousands whose labors have perished. The eagle and the vulture reign supreme, and build their nests upon the branches of huge trees whose evergreen hide the mournful skeleton; serpents, with fiery eyes and forked tongues, have their dens in magnificent halls where "joy and festivity" formerly resounded. Great spiders indigenous to the tropics weave their webs among the ruins, and hang upon their silken cords from the dome of the palace; birds of song and beautiful plumage warble their matin chorus, perched upon the desolate walls. Time has marked her for a victim—her glory has departed.

"She lives, but in the tale of other times.
Her proud pavilions are the hermit's home,
And her colonades, her public walks,
Now echo faintly to the pilgrims feet,
Who comes to muse in solitude, and trace
Through the rank moss revealed, her honored dust.
And not to her alone has fate confined
The doom of ruin—cities numberless,
Tyre, Sydon, Carthage, Babylon and Troy,
And rich Phœnicia; all are blotted out,
Half razed from memory, and their name
And being in dispute !"

RUINS AT COPAN.

One of the most interesting and wonderful ruins fonnd in Central America is located in the western part of Honduras and about one hundred miles from the ocean. This ancient city has been denominated Copan. Like Palenque, Copan is buried in an almost impenetrable and pathless forest, where it has stood in mournful solitude scores of centuries, as its present appearance indicates a greater antiquity than most of the southern ruins. Notwithstanding the great amount of territory occupied by the Aztecs and their numerous inhabitants, Copan was unknown to them at the time of the fall of their empire. About forty years after the conquest, the Spaniards, while exploring the river Copan, discovered the ruined city on the left bank of the river, and extending along its borders for many miles.

These grand and imposing ruins were examined by Mr. Stephens while traveling in Central America. In describing them he says: "We saw directly opposite a stone wall from sixty to ninety feet high, with furze growing out of the top, running north and south along the river 624 feet, in some places fallen, in others entire." The colossal wall, like most of the great structures in Mexico and Central America, stood upon an elevated foundation and terraced upon either side.— This imposing edifice was constructed of hewn stone laid in cement which has stood the ravages of time for thousands of years. Near this ruin stood a stone column fourteen feet high and about three feet square. Two of its sides were enriched with ornaments beautifully carved, and for any age or country would do credit to a master hand. Fourteen other columns were also discovered, all of which were elegantly sculptured and beautifully designed. This enormous building must have been nearly forty rods in length, and surrounded by a wide terrace that stood parallel with the river and sufficiently elevated to command a view of its waters for many miles.

Monarchs of the wood, gigantic in size, are growing upon the terraced foundation, some of which are between twenty and thirty feet in circumfrence. These trees are very old, and for hundreds of years have bowed their heads to none but the hurricane, and kissed the breath of the gentle breeze when freighted with the morning dews.

This ancient city, located in this dark and tangled wilderness, is rich with wonderful relics which show that architecture, sculpture and the various arts practiced by a cultivated people were once carried on where now is a dense and desolate wilderness. Besides the huge building many pyramidal structures exist which are mentioned by Mr. Stephens who claims that "extensive explorations are impossible unless one shall first clear away the forest and burn up the trees."

Palacois, who visited Copan nearly three hundred years ago, described it as a large city. Among the wonders that attracted his attention was "an enormous eagle carved upon stone which bore a square shield upon its breast covered with undecipherable characters." He also discovered among the ruins a "stone giant" with one arm broken, and also a "stone cross."

This great building is grey with age; thousands of summer suns have shed their golden rays upon this structure which has long mocked the daylight with its huge proportions. It stood before the Coliseum at Rome had an existence, and very likely was anterior to the great temple of Belus at Babylon, or the massive walls which form the colossal rampart of China. The ruined walls of this immense structure far exceed those of the great temple at Baalbec, and for size its foundation is hardly excelled by that of any building in the world.

If the crumbling ruins when seen by Mr. Stephens were from sixty to ninety feet high, it is apparent it once was far more towering, for it was surrounded with prostrate pillars and large ornamented stones which once adorned its cornice.

MITLA.

About thirty-six miles from the famed city of Oxaca is located the ancient city of Mitla, which is situated in a wide and extensive valley. This valley, where once teemed a great and tumultuous throng, now reigns in the solitude of death. A large area is covered with the remains of buildings scattered in great confusion. Mr. Charnay visited these ruins in Mitla in 1860, and describes four standing edifices, which "were erected with lavish magnificence and combined the solidity of the works of Egypt with the elegance of those of Greece." Much of the architecture and masonry resembles that of Palenque, though for elegance of design and elaborate workmanship, it is far superior. Mr. Charnay in describing the grandeur and beauty of the decorations, places them with the magnificent monuments of Rome in the days of her greatest prosperity and glory. Dupaix places these works the highest among the known works of architecture now in existence. According to his description much of the external surface of the walls are inlaid with various colored stones and cemented on stucco in such a manner as to imitate beautiful gradations in painting. These great and elegant structures present the most marvelous and interesting designs and elaborate workmanship ever discovered among the greatest efforts of human skill. The labor which must have been performed upon those inlaid decorations, which extends for hundreds of feet below the cornice and the lower stories must have been immense, and it would seem to have required the combined labor of hundreds of artisans for many years.

It is the opinion of archæologists that Milta and many other cities in Mexico and Central America have been occupied by a people far below the builders in architecture and civilization, and that that people took possession centuries subsequent to the abandonment by those who built them. On that subject Mr. Charnay says: "It is a bewildering maze of courts and buildings with facings ornamented with mosaics in relief, of the purest design; but under the projections are found traces of paintings wholly primitive in style, in which the right line is not even respected. These are rude figures of idols and meandering lines that have no significance. Similar paintings appear, with the same imperfection, on every great edifice in places which have allowed them shelter against the ravages of time These rude designs, associated with palaces so correct in architecture and so ornamented with panels of mosaic of such marvelous workmanship, put strange thoughts in the mind. To find the explanation of the phenomenon, must we not suppose these places were occupied by a race less advanced in civilizatiou than their first builders?"

Fig. 32—Ancient Ruins at Mitla.

Westerly from these ancient ruins, at a distance of about two miles, stands a strange and wonderful structure which has been denominated the "castle of Mitla." This curious edifice is located upon the apex of a precipitous hill which is composed of rock and is accessi-

ble only on the side bordering the valley.   The level
summit of this eminence is entirely surrounded by a
colossal wall composed of hewn stone twenty-one feet
thick by eighteen high.   "The east side is flanked with
double walls " and within the walls are several buildings now in ruins.   This castle is placed in a position
that overlooks the city and an extensive section of the
surrounding country.   Mr. Baldwin, in his "Ancient
America," gives a brief sketch of this remarkable building, and leaves it without any attempt at explanation.
But it is obvious that such an expensive and imposing
edifice and so massive a structure must have had an
ethnological significance.   A further observation sug-
gests that it had either a priestly or military func-
tion.   The circumstance of several buildings being
placed within the enclosure would indicate that it was
continually occupied, and that the internal structures
were designed to afford protection for a great number
of persons, whose business it was to perform a duty
for which they were appointed.   The fact that the
castle was placed upon a high hill and built with such
remarkable strength as to defy almost any force that
we can conceive could have been brought to bear upon it, would incline us to believe that it was designed
for military purpose.   From the best imformation we
can gather it seems to stands above a gorge made by
the waters of the river and flanked on either side by
precipitous hills, and that the gorge would be the only
place well calculated for the passage of any army desirous of reaching the city from a certain direction ;

and that this impegnable rampart was designed to stop the approach of the enemy in their movements towards the city. Castles of a similar character were formerly constructed and used for military purposes in Europe before the invention of gunpowder. When Jerusalem was taken by Titus, the soldiers hurled their deadly missiles upon the heads of the enemy from the massive walls.

In the vicinity of Mecamecan, Capatin Dupaix discovered a solitary granite rock which had been formed into a pyramid, by human labor, that was undoubtedly devoted to astronomical purposes. On the summit was constructed a horizontal plane, where a number of hieroglyphical figures are sculptured. On the south side of this level summit is the figure of a man holding to his eye an opticle instrument directed towards the heavens, and near his feet are "six celestial signs carved on the surface of a rock." That this singular work which has attracted so much attention has an astronomical significance is now regarded as well estabished. "Telescope tubes" of artistic workmanship, have also been found among relics disclosed in sepulchral mounds in the United States. In one of the tombs explored in ancient Peru, the figure of a man wrought in silver and holding a teloscopic tube, has been discovered. That these ancient people were considerably advanced in many of the sciences is very apparent. Many of their works show the highest of mechanichal skill where a knowledge of geometry was indispensible. The terraced foundations of these magnificent buildings so widely scattered over a vast territory, attest beyond a question that they had an extensive knowledge of mathematics.

UXMAL

In Yucatan are the ruins of an ancient city known as Uxmal which attracted the attention of the Spanish people immediately after the conquest.

Fig. 33. Magnificant structure at Uxmal.

This ancient city covers a vast amount of territory but is now mostly in ruins. One important structure now remains which attracts much attention on account of its great size and elaborate workmanship. It is 320 feet long and constructed of hewn stone. This wonderful structure is shown in Fig. 33

RUINS IN SAN SALVADOR, HONDURAS AND GUATAMALA.

In the southern part of Central America which includes a portion of San Salvador, Honduras and Guatamala, the ancient ruins can be counted by the hundreds. Mr. Squier, one of the most prominent explorers, discovered many which presented the remains of subterranean galleries; circular towers and "vast edifices," all constructed of hewn stone. Near the northern shore of Lake Nicaragua, among the ruins, large quantities of pottery which display wonderful skill in its manufacture, and equals the most beautiful specimens found in Mexico or South America, has been discovered.

On a mountain in San Salvador, near the town of Mecallo, are the remains of a large city, which presents streets and various subterraneous passages, together with magnificent structures, many of which are not inferior to those found among the ruins of Copan or Palenque. A Spanish writer referring to these ruins says: "The buildings were constructed of thin stones or a species of slate united by a kind of cement, which in appearance resembles melted lead." The most interesting and wonderful work found here, is a temple devoted no doubt to religious worship. In the vicinity of the entrance and on the archway are representations of the sun and moon, carved in the solid rock, by the hand of a master workman. Hieroglyphical figures are found in the interior, sculptured in a style which shows the most remarkable skill displayed in any age or country. Among the works found in this consecrated temple is the figure of an animal, of great size, sculptured on a rock.

North of the dense wilderness in the State of Yucatan, the country is literally strewn with ancient and solitary ruins, which attest beyond a doubt a vast and wonderful population by which that tropical country was once inhabited.

Mr. Stephens, in his interesting work on Yucatán, says that he discovered "forty-four ruined cities," which time and the "elements are hastening to utter destruction." At Mayapan which was once an ancient capital, is located a great mound that attracts the attention of the antiquarian as he approaches the city.

This imposing structure is partially hidden from observation by huge trees, by which it is surrounded, and the forest that entirely covers its surface is raised far above those located on the level plain. This mound is sixty feet in height and one hundred feet square at the base. The summit is approached by four stairways which lead to within six feet, which last distance is reached by a single set of stairs. This singular work was very likely a prominent temple which had a religious significance. "Subterranean chambers" have been discovered within its walls.

The ancient buildings which formerly comprised the city of Mayapan are mostly in ruins. The only one that now remains is circular. It stands on a pyramidal foundation which is located about thirty-five feet above the surrounding plane. On the south side of this pyramidal structure is a wide terrace that projects from the base on which is a double row of columns about eight feet apart. This remarkable edifice was built in a style similar to those found in Copan and Palenque, and must have been uncommonly firm to have withstood the action of the elements, while those which surrounded it are reduced to heaps of ruins.

Mayapan, like most of the great cities in Mexico and Central America, has been several times rebuilt. In 1620 the modern city was destroyed by the Spanish army, who sacrificed every element of humanity in their search for gold and precious stones. What appears the most wonderful to antiquarians is the fact that nearly all of the ruined cities, show that the first builders were far superior in civilization and the arts, to those that came after them. All of the old works show those builders as far advanced in architecture as any other people that ever lived upon the earth.

At Ake, in Yucatan, stands a great mound that presents an uncommon interest to the antiquarian. On the summit of this truncated tumulus, stands thirty-six columns or shafts, which are placed in three parallel

rows.   These shafts average about fifteen feet in height
and about four feet square.   For what purpose this
great work was constructed is not only a mystery to
us, but is unknown to the present inhabitants who occupy the territory where once stood an opulent city,
one of the largest found among the ruins in Central
America.

On this peninsula there is, and no doubt always has
been, a great scarcity of water.   Artificial ponds which
present wonderful skill in their mode of construction
can now be seen in a good state of preservation.   These
are very old and were probably built when those wonderful structures were erected which have so astonished the archæologists while wandering amid these ruins.
The reservoirs are composed of hewn stone laid in cement, and at the bottom were large cavities impervious to water.   Great numbers of such reservoirs were
constructed for the use of that once imposing and populous city.   Subteraneous reservoirs that led to living
pools of water deep down in the rock were reached at
an enormous expense. One of these at Galal is 450 feet
below the surface and the circuitous passage which
reaches it is 1,400 feet in length.   This winding tunnel is so constructed that the thirsty traveler can reach
it with ordinary toil.   "Necessity is the mother of invention," and a great necessity penetrated the earth to
the depth of 1,400 feet to this fountain of living water.

Nearly all parts of Central America adapted to the
existence of man is a vast field of ruins.   Hundreds of
large cities which have been discovered since the con-

quest, lay scattered in great confusion, where once lived a people who were highly cultivated in the arts and
many of the sciences. What now remains of their

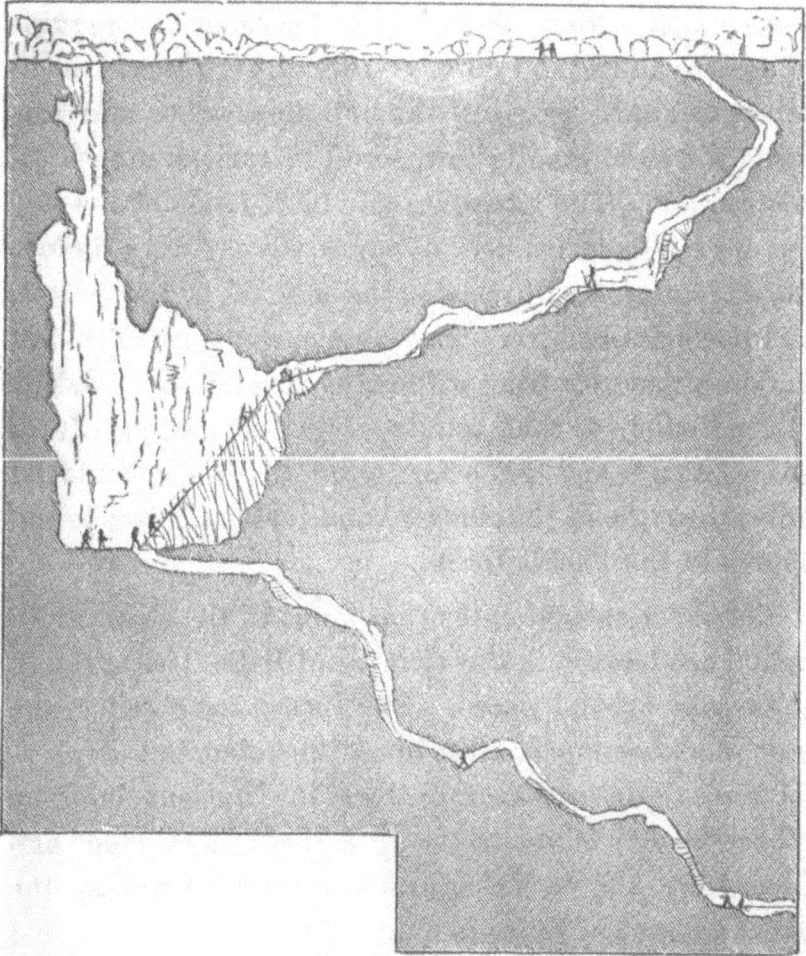

Fig. 34—Great Reservoir 1,400 feet below the surface.

works carry unmistakable evidence of their civilization and greatness, but the scattered fragments of

great temples, the solitary wilderness within which stand majestic structures now crumbling to ruin, shows the utter feebleness of man and the littleness of his greatest works when touched with the revolving wheels of time.

As we pass from Central to South America the ruins of ancient towns and cities continually present themselves to view. Some of the most marvelous and majestic works of man lay scattered in various stages of destruction. The mountain and table lands of the Andes are a wonderful field of ruins which, ages before the advent of the Incas, was the seat of a prosperous and powerful empire. The most ancient ruins found in South America bear witness of at least two periods in the history of that country. These extend from Lake Titicaca and Quito for more than five hundred miles throughout the elevated and fertile regions to the shores of the Pacific Ocean.

The most ancient ruins that point to the older civilization are located in the vicinity of Lake Titicaca and the region around Cuzco. The Incas, long after the city was abandoned, built upon the ruins, but displayed less skill in architecture than the ancient builders whose history is lost in the ages that have long ago passed into the dark shadows of eternity. Among the great amount of relics found in Cuzco may be included a lunar calendar made of fine gold, which from the engraving found upon it is supposed to have an astronomical significance. What has been found in Mexico.

South and Central America makes it apparant that the people who built so many great cities and such gigantic structures had a knowledge of astronomy. In 1791, near the present City of Mexico, was found a great astronomical calendar twenty-seven feet in circumference and weighing twenty-four tons. This remarkable stone had engraven upon it hieroglyphics signifying the divisions of time and the twelve signs of the Zoliac and the positions of some of the heavenly bodies. This stone was buried in the ground, together with the religious symbols and books of the Mexicans, by the orders of the Dominican and Franciscan "fanatics, whose learning and religion consisted of ignorance and bigotry," for they were fearful that the religious views of the Mexicans might interfere with their monkish superstitions.

Las Casas, one of the Catholic missionaries, refers to the destruction of a vast collection of books and writings which were made on paper. With regard to those books he says: "They recorded the history of kings and the modes of their election and succession; of their labors, actions, wars and memorable deeds, good and bad; of the virtuous men or heroes of former days, their great deeds, the wars they had waged and how they distinguished themselves; who had been the earliest settlers, what had been their ancient customs, their triumphs and defeats. They knew, in fact, whatever portained to history, and were able to give an account of all past events. These chroniclers had likewise to calculate the days, months and years; and

though they had no writing like ours, they had their
symbols and characters through which they understood
everything ; and they had great books which were com-

Fig. 35—Picture Writing found among the Ruins, Engraved on Stone.

posed with such ingenuity and art that our characters
were really of no great assistance to them. Our priests
have seen those books, and I myself have seen them

likewise, though many were burned at the instigation of the monks, who were afraid they might impede the work of conversion."

Nothwithstanding, these bigoted priests undertook to destroy every vestage of the religious and historical writings found among this conquered people, a few were saved from the flames. Several persons have claimed to translate some of these ancient books into different languages. Brasseur de Bourbourg has translated one of the translations of a Spanish writer into French. If these translations are correct, it shows that those people had high and noble conceptions of a God, and his mode of governing the world ; it shows they admired virtue and everything noble in human nature, and regarded their domestic institutions of the most sacred character, where they could enjoy in full fruition the joys and pleasures peculiar to happy homes.

GREAT ROADS IN SOUTH AMERICA.

Among the most wonderful works in South America are the great roads, whose beds are twenty-five feet in width, and made permanent with pulverized stone, "mixed with lime and bituminous cement." These roads were protected on either side with a strong wall several feet in thickness, and show in unmistakable language the great skill and vast labor performed by those who built them. The ancient roads were built along the sides of great mountains and through dismal chasms bordering the sierras; rocky precipices and deep ravines offered no resistance to the daring skill of those ancient engineers. Deep gullies, so numerous on the sides of towering hills, were filled up with "solid masonry." Mr. Baldwin in referring to these marvelous works says: "The builders of our Pacific railroad, with their superior skill and mechanical appliances, might reasonably shrink from the cost and difficulties of such a work as this.

"Extending from one degree north of Quito to Cuzco, and from Cuzco to Chili, it was as long as the two Pacific railroads, and its wild route among the mountains was far more difficult."

Sarmiento describing it said: "It seems to me that if the Emperor (Charles V.) should see fit to order the construction of another road like that which leads from Quito to Cuzco, or that which from Cuzco goes towards Chili, I certainly think he would not be able to make it, with all his power."

This nineteenth century is remarkable for the great discoveries that have been made, not only in the arts. but in many of the sciences. Astronomers have discovered new and unknown fields in the heavens, and analyzed the elements of revolving worlds. Geology as a science has been born, and revealed the first and lowest forms of animal life, that sported beneath the oceans' waves, and traced the thousands of new creations up to man. Chemistry has made rapid strides and poured a flood of light upon the dark and hidden mysteries of nature. Railroads and steamboats have been invented which convey millions of men, women and children across the ocean and over the rocky mountains, reposing in elegant palaces, protected from wind and storm. The magnetic telegraph has brought the civilized world within speaking distance, and carried intelligence among the coral-decked chambers of the mighty deep. Agriculturists have added new and numerous inventions that have abridged the time of human toil and multiplied its results.

Notwithstanding the age in which we live has been so prolific in new inventions and the arts, it is very likely that much which would be new to us, has been buried with the millions of ancient inhabitants who have left such wonderful ruins in Europe and America. There is "nothing new under the sun." Humboldt in his "cosmos" says that the Chinese had magnetic carriages with which to guide themselves across the great plains of Tartary, one thousand years before our era, on the principle of the compass. The prototype of the steam engine has been traced to Hero. of Alexandria. Movable types were used by the Romans two thousand years ago, "to mark their pottery and endorse their books." Mr. Layard discovered, in the ruins of Nineveh, a convex lens composed of rock crystal, which was considered by Sir D. Brewster an optical lens and the origin of the microscope. The principle of the sterescope was known to Euclid fifteen hundred years ago. The Thames tunnel was anticipated by the one under the Euphrates at Babylon, and the ancient Egyptians had a Suez canal.

CHAPTER XIV.

THE GREAT AGE OF THE MOUNDS AND THE PROBABLE ANTIQUITY OF THEIR BUILDERS.

The time when the Mound Builders occupied so large a portion of North America as their works would indicate has been a subject of much speculation. Some have put forth and advocated the doctrine, as formerly observed, that the mounds were constructed by none other than the various Indian tribes. This class of antiquarians claim that the mounds and fortifications so extensive in the fertile sections of various states are of comparatively recent origin, and antedate the present Indian but a few hundred years. This theory, which formerly seemed very plausible, has now ceased to be advocated. A more thorough investigation of the ancient tumuli point to a people far more advanced in civilization than any the Indian nations. The Indian has but little affinity for the blessings of civilization, for by nature he is a wandering savage and prefers for his home the solitude of the wilderness and the mountains. The rich valleys of Ohio and Mississippi and the garden lands found in the west, which were once cultivated by the Mound Builders, became

a wilderness and was but little disturbed by the Indians since left to the wild influence of nature.

That the Mound Builders were once far more numerous than all of the Indian tribes is apparent from
the vast amount of relics so widely scattered. In Western New York the territory is rich with the relics of
the Mound Builders. On every farm whether located
in the valleys or on the hills, fragments of pottery,
gorgets, stone axes, flint arrow-heads and frequently
implements of copper are found. Within the last twenty years I have consulted more than a hundred farmers and they invaribly claim to have found numerous
relics wrought by the hands of an ancient people.

Some of the hills bordering the valley of the Allegany have been cultivated to some extent, and on these
hill lands as well as in the valleys the plow brings up
the evidences of ancient occupation,

It has been suggested by several archæologists that
the circumstance of the mounds and fortifications being never found on the lowest terraces formed by the
the rivers and tributary streams, points to the great
antiquity of those works. It is a fact well known to
every person who has examined the subject that all
streams of any considerable magnitude show several
different terraces. which unmistakably mark as many
different eras in their subsidence since their channels
were marked out by the rush of waters. The first and
lowest terrace bordering the lakes, where now are
splendid villages and fertile farms, show no marks of
an ancient occupation, hence they infer that that ter-

race was formed subsequent to the construction of any of the ancient works.

But the time these ancient works were constructed will never be known. The tongue which could reveal the wreck of empires, the character of the people who built those structures, from whence they came and whither they went, is silent and long ages ago mingled with the dust. The tombs alone must tell the story.

That many of the mounds were completed thousands of years ago is beyond a doubt. The skeletons are mostly found in the last stages of decay, notwithstanding they have been so wonderfully protected. The sepulchral mounds are so constructed that the location of the dead is impervious to water. The great mound at Grave Creek when first explored was nearly free from dampness and the soil which surrounded the dead had been protected from the action of water since the great work was completed. Now it is evident that when the bones are thus protected and lay undisturbed the earthy portions will change but a little in many centuries. Those people, from what we can discover, desired to preserve the remains of their distinguished personages, and well they have done it, for their bones can be detected milleniums after ours are past discovery.

Many antiquarians have been misled with regard to the great antiquity of the sepulchral tumuli from the finding of human skeletons near the surface of the mounds in a good state of preservation. These are the skeletons of some of the Indian tribes who had been in the habit of burying their dead over the ancient remains.

The bones of the Mound Builders are always found near the surface of the surrounding soil while those of the Indians are deposited near the apex of the mound.

The remarkable antiquity of the mounds is also shown by their being buried in the midst of great and extensive forests which was cultivated land when they were constructed. Trees of great size and which must have begun their life many centuries ago are frequently found on the surface of the mounds.

The circumstance of the ancient works being found in a forest, the age of which is over a thousand years, fails to disclose their antiquity. It is very probable that many generations of the forest monarchs have gone to decay since that unknown people mourned for some distinguished character who had won a monument wrought by the labor of hundreds of men for many years.

Whether the builders of these great works were the descendants of the first inhabitants of America is also shrouded in darkness. That this country was occupied for ages prior to the construction of the mounds is apparent, for there is a vast amount of territory in which evidences of previous occupancy now exist.

Those inhabitants evidently explored every part of the country, even to the shores of Lake Superior where they found a metal which they considered of uncommon value and which they distributed in almost every part of the United States long before the great tumuli were

constructed. It is very apparent that there were inhabitants of this country many thousand years before
they discovered the great copper mines, the products of
which they spent so much labor to obtain. As the section of country around Lake Superior would be unfavorable to a large settlement of that people on account
of the poverty of the soil and the extreme cold during
the winter months, it is very probable that after they
had arrived to some moderate state of civilization they
made it a business to explore the various portions of
the country in search of materials for the construction
of their implements of war and husbandry.

The first discovery of copper in the regions of Lake
Superior was undoubtedly accidental and was found
near the surface, as large masses of native copper are
frequently discovered, one of which was discovered by
Schoolcraft that weighed about two thousand pounds.
After the first discovery was made and the great value
of the metal determined, they commenced the business
of mining on a scale which, as shown by their works,
was quite extensive. Keweenaw Point, a projection of
land which extends into Lake Superior to the distance
of eighty miles and is about forty miles in width at its
junction with the main land, is the territory where their
most prominent works have been disclosed. This great
territory which is far more extensive than that which
has been worked by the white man has all been thoroughly explored. The ancient works are numerous in
almost every direction.

The excavations made by the Mound Builders form a

part of the aged wilderness which was growing at the time the modern mining commenced. The vast amount of time which has passed away since those mines were worked has filled those remarkable excavations with debris to the level of the surrounding soil, and planted upon their surface several generations of forests which has been attested by the remains of large and ancient trees "lying across the pits."

That these copper mines were not discovered until long after the Mound Builders appeared upon the platform of existence is very probable, and that those mines were worked previous to the construction of the great mounds (which have been frequently referred to) is also apparent. The copper implements buried with the dead, whether found in Western New York or in the valley of the Ohio or Mississippi, were all brought from Lake Superior. From the appearance of the great amount of relics found it is evident they worked none but native copper, and were not sufficiently versed in chemistry to obtain it when in combination with other substances.

Notwithstanding these ancient miners performed immense labor in gathering the precious metal on the shores of Lake Superior, they have left no evidence behind them that they had permanent settlements there or cultivated the soil. Copper was the great incentive that led them from their homes located on the fertile lands bordering the southern rivers, through the dark and forbidding wilderness surrounding the great lake. That they transported some of their provisions from the south is very probable, but that fish and the flesh of wild animals constituted their principal food is beyond a doubt.

The theory that is now generally adopted is that they spent their summers in collecting copper and instead of braving the rigors of a northern winter left for a more congenial clime. It would seem impossible for so great a number of people as worked those northern mines to sustain themselves there during the winter months. The lake from which they derived a large portion of their subsistence would be clothed with snow and ice. The sea-fowl, guided by a wonderful instinct, would leave the ice-bound waters and fly to more southern climes, where they could bask in the liquid element so congenial to their nature. Wild animals whose flesh is adapted to food march in hordes from the northern wilderness at the approach of winter. But at the opening of spring, when the melted snow leaves bare the brown earth, enormous flights of birds, beasts and fowl would enliven the dismal scene during the months of summer to pay tribute to man's many wants, "to appease his hunger" and clothe his body or to gratify his inordinate thirst for gain.

When these great copper mines were explored in 1845 and the extensive ancient works discovered, archæologists hoped they had a key to unlock some of the strange and wonderful secrets which surround the ancient Mound Builders. One important fact already has been revealed, and that is where they obtained the great amount of copper so widely distributed in the sepulchral mounds. The copper from Lake Superior can be detected wherever found, especially that which has never been smelted, for that which is taken from the veins in a native state invariably discloses specks of silver mingled with the metal; hence whenever native copper has been found in the mounds thus characterized it was mined at Lake Superior. Copper from those mines has been disclosed in nearly all of the great mounds as far south as the Gulf of Mexico. The great apparent antiquity of these mines has led many to believe that the Mound Builders were much more ancient than it was formerly supposed.

Among those who advocate the doctrine that the mounds were recently constructed is a class that still adhere to the Indian theory and claim that during the past few hundred years the Mound Builders have degenerated and lost their disposition and ability to construct such great and imposing tumuli that we find so widely scattered in the fertile borders of prominent rivers.

That war and superstition running through numerous generations may reduce civilization of man and sink him to a state of barbarism, is attested by the history of the past, but the descendants of a degenerate people always retain some traces of their lost civilization which will be transmitted to succeeding generations for thousands of years. The low and barbarous condition of the Indians found occupying the American Continent at the time of its discovery by Europeans would indicate that they had never attained to any higher de-

gree of civilization than they presented when first discovered.   The fact that the Indians had traditions of
their ancestors who occupied this country many centuries in the past, and that they had no knowledge of
the Mound Builders only by the works which they had
left behind, would add to instead of diminish the probability of their great antiquity.   That these unknown
people lived and occupied this country many thousand
years ago, and that they had attained a much higher
civilization than any of the Indian nations, needs no ad-
ditional evidence, but to measure the time of their existence, either by centuries or milleniums, is beyond the
reach of history or traditionary lore.   Many of their
works are now obliterated while others lay hid in the
dark and dismal forest, waiting for some savant to appear and interpret their mysteries.

THE CIVILIZATION AND ANTIQUITY OF THE ANCIENT IN-
HABITANTS OF CENTRAL AMERICA.

The remarkable ancient civilization that ages ago
existed in Central America has been a subject of much
attention among archæologists, theologians and histori-
ans   Since the Conquest, Mexico has been extensively
explored, not only by the Spanish conquerors, but by
savants from every country in the civilized world. Hundreds of theories have been advanced, nearly all taking
the ground in the outset that any civilization found in
America must from necessity have been imported from
the Old World.

THE ISRAELITISH DOCTRINE.

The theory that the civilization of the ancient inhabtants of Central America come from the "lost tribes of Israel" was one of the first adopted. The Spanish priests who were sent out to Mexico immediately after the Conquest to christianize that pagan people, advocated the doctrine that "the gospel was preached in America by St. Thomas." Notwithstanding this theory is the most "unwarranted and absurd" of any which have been presented, some of the prominent writers on the antiquity of Mexico and Central America have given volumes in its favor. Lord Kingsborough adopted this theory and labored extensively to sustain it. The finding of the sculptured cross in the ruins of Palenque was formerly considered almost a demonstration that that symbol belonged exclusively to christianity, and that the builders of those ancient cities lived and occupied that country subsequent to the coming of Christ. But as has been shown, the cross is not exclusively a christian emblem, and can be traced to the oriental world, where it was adopted as a religious symbol by the Phœnicians unknown centuries before the advent of Christianity.

With regard to this theory of the "lost tribes." Mr. Baldwin says: "According to the truly monkish theory, the 'lost ten tribes of Israel' left Palestine, Syria, Assyria, or whatever country they dwelt in at the time, traversed the whole extent of Asia, crossed over into America at Behring's Strait, went down the Pacific coast and established a wonderful civilization in that part of the continent where the great ruins are found. The kingdom of the tribes was destroyed not long previous to the year 700 B. C. How many years are allowed after the escape from captivity for this unparalleled journey, has not yet been asertained. But if such a journey had been possible, it would have resulted in utter barbarism rather than any notable phase of civilized life. Even the Jews who remained faithful to Moses, although important on account of their scriptures and their religion, were not remarkable for civilization. They were incapable of building their own temple without aid from the Tyrians. Moreover, there is not either a fact, a suggestion or a circumstance of any kind to show that the 'lost ten tribes' ever left the countries of Southwestern Asia, where they dwelt after the destruction of their kingdom. They were 'lost' to the Jewish nation, because they rebelled, apostatised. and after their subjugation by the Assyrians in 721 B. C., were to a great extent absorbed by other peoples in that part of Asis. Some of them probably were still in Palestine when Christ appeared. This wild notion called a theory scarcely deserves so much attention. It is a lunatic fancy, possible only to men of a certain class which in our time does not multiply."

"THE 'PHŒNICIAN' DOCTRINE."

The theory that is now maintained by some is that Central America was originally peopled by the Phœnicians. It appears to be well authenticated that the ancient Phœnicians were a very intelligent and civilized people, who devoted much attention to navigating the ocean, and that they displayed uncommon skill in their maratime affairs. It is said they traversed all the seas, discovered new countries, and introduced their knowledge and civilization far beyond the "pillars of Hercules."

That the ancient Phœnicians, as they are called, in pre-historic times, lived and flourished in the vicinity of the Mediterranean and around the Strait of Gibralter, "before Tyre and Sidon were built," is very probable. That the ocean was navigated ages ago by the the pre-historic inhabitants of Northern Asia is not very difficult to believe.

Since man appeared upon the earth great changes have taken place with regard to his civilization and his works. The builders of the great pyramids were a people far advanced in geometrical and (no doubt) astronomical knowledge, and it is highly probable that they and their ancestors navigated the ocean in ages past and that great ships laden with the products of South and Central America were landed near the Ægian Sea, at Alexandria in Egypt and numerous places on the continent of Europe. When we cast our eyes upon the mighty pyramids we see some of the grandest efforts of human genius displayed in those magnificent works, and are reminded of the language of the historian, who says: "I was bewildered and said in my thoughts this is not the work of man; none but an Almighty power could have built this pyramid."

I trust I shall not be accused with credulity when I believe that the builders of the pyramids and of Baalbec and Palmyra navigated the ocean with ships and had commercial relations with the cities of Palenque, Copan and other great cities in Central and South America.

It is said that the ancient Mexicans have retained a tradition that their ancestors crossed the ocean in ships. Diodorus Siculus, a Greek historian who flourished in the fourth century, gives an account of a Tyrian ship which crossed the ocean and found a country noted for the magnificence of its cities and the grandeur of its inhabitants. But the writings of this historian, though very extensive, are unreliable and often fabulous.

It is argued by the advocates of the theory that these ancient inhabitants of Mexico and Central America originated in Phœnicia or some part of Southern Asia, for the reason that the symbols found in their ancient tombs are similar to those used by the Phœnicians.

Notwithstanding there may be some similarity in the "symbolic devices" found in many of the American mausolea, it fails to establish the theory. Among all of the pre-historic nations whose tombs have been explored by the modern inhabitants both of Europe and America, similar relics have been found. Nearly all the pagan nations that lived in pre-historic times were worshipers of Baal. The sun is one of the most prominent objects in the heavens; he changes the darkness of midnight into the light of day; prepares the cold earth for the development of vegetable life; converts the waters of the ocean, lakes, rivers and stagnant pools to vapor, which falls to the earth in shape of rain to vivify the thirsty plants designed for food; rarifies the air which brings health to human habitations, and causes the gentle zephyrs of heaven which are continually fanning the human brow with a touch of delight; and then again throws the atmosphere out of balance, from which comes the awful hurricane or "roaring tornado," which levels human habitations in its course and lays prostrate the mountain forest.

As heretofore stated, the most prominent element in human nature is to worship, and as the sun is the most conspicuous object and rides majestically through the heavens dispensing the blessings of light and heat upon the earth, it is not strange that he should become an object of worship to a people unfamiliar with the laws that govern the universe. Coleridge on seeing the sun rise on Mount Blanc is touched with veneration:

"Rise, O, ever rise !
Rise like a cloud of incense from the earth !
Thou kingly spirit throned among the hills,
Thou dread embassador from earth to heaven.
Great Hierarch ! tell thou the silent sky,
And tell the stars, and tell yon rising sun,
Earth with her thousand voices praises God."

The class of antiquarians who deny the Phœnician theory, do so from the fact that the mode of building adopted by the ancient inhabitants of Central America bears no resemblance to that of ancient Phœnicia. The former placed their great and expensive structures upon pyramidal foundations and cased them with hewn stones, a mode which has never been practiced on the eastern continent by any of the nations of antiquity. The style of writing practiced by the builders of Palenque and other great cities in Central America bears no resemblance to that of Phœnicia or Palestine. Notwithstanding they "have been annexed and sometimes separated from the jurisdiction of Syria, yet the civilized world has received their religion from Palestine and their letters from Phœnicia."

THE LOST ISLAND.

Brasseur de Bourburg, who has investigated the writings and symbols found among the ruins of Palenque and other ancient cities in Mexico and Central America, advocates the doctrine that the civilization of Ancient America sprung from the "lost island of Atlantis." He claims that the Western Continent once extended from New Granada into the Atlantic ocean, and that from some great convulsion of nature it was sunk beneath the waves ; and that "the Canary, Maderia and Western Islands may be remains of this portion of it."

That the ancient orientals had a tradition that the lost island of Atlantis was once inhabited, and was the seat of a remarkable civilization and a powerful empire, is apparent from the story being "preserved in the annals of Egypt." In Cousin's translation of Plato's history occurs the following story :

"Among the great deeds of Athens of which recollection is preserved in our books, there is one which

should be preserved above all others. Our books tell
that the Athenians destroyed an army which came
across the Atlantic Sea, and insolently invaded Europe
and Asia, for this sea was then navigable, and beyond
the strait where you place the Pillars of Hercules there
was an Island larger than Asia Minor and Lybia combined. From this island one could pass easily to the
other islands, and from these to the continent which
lies around the interior sea. The sea on this side of
the strait (the Mediterranean) of which we speak resembles a harbor with a narrow entrance, but there is
a genuine sea and the land which surrounds it is a
veritable continent. In the island of Atlantis reigned
three kings with great and marvelous power. They
had under their dominion the whole of Atlantis, several other islands and some parts of the continent. At
one time their power extended into Libya, and in Europe as far as Tyrrhenia, and uniting their whole force
they sought to destroy our countries at a blow, but
their defeat stopped the invasion and gave indepen-
dance to all countries on this side of the Pillars of Hercules. Afterwards, in one day and one fatal night,
there came mighty earthquakes and inundations which
engulfed that warlike people. Atlantis disappeared
beneath the sea, and then that sea became inaccessable
so that navigation on it ceased on account of the quantity of mud which the engulfed island left in its
place."

Since the appearance of man upon the earth it is
not improbable that a long peninsula extended from

Central America far into the Atlantic, and the West India. Canary, Madeira and Western Islands may be the remains which mark out its course. If such a portion of land as the "lost Atlantis" stood above the ocean far back in the past, the above islands were towering mountains which defied the earthquake that overthrew the cities and their inhabitants.

Brasseur de Bourburg claims that there were traditions among the ancient Europeans that such a country once existed and that it was extensively inhabited by a race of civilized men who constructed cities ; but that "the land was shaken by frightful earthquakes and the waves of the sea combined with volcanic fires to overwhelm and engulf it." He also claims that a succession of fearful convulsions swept away the land, and those who survived the terrible cataclysm were saved upon the mountains where they stood above the waters.

Advocates of the Atlantic theory also claim that symbols have been found in Central America which point to the "lost Atlantis," and some of those who survived the terrible catastrophe were brought away in ships and related the story of their misfortunes.

That terrible earthquakes accompanied with volcanic fires have made islands in the ocean and inundated cities is a historical fact. In 1868 a fearful earthquake occurred in Peru which came with an undulating wave and shook the country like a mighty avalanche. The city of Ariquipa, which was located forty miles from the ocean, was destroyed in five minutes with a great

number of its 44,000 inhabitants   On the sea coast
was situated Iquique and Arica ; both were destroyed
by the shocks and overwhelmed by a tremendous wave.
The ocean thus took up the vibrations of the land, and
waves of great violence were put in motion which
rolled not only along the coast but away from it with
a velocity in the deep ocean of not less than 400 miles
an hour.   The great wave—for one was of greater volume than the others—has been estimated at upwards
of 200 miles in breadth, with a length along its curved
crest of 8,000 miles.   This rolled into the harbor of
Yokohama in Japan, 10,500 miles distant, and was felt
at Port Fairy in South Victoria, distant one-half the
earth's circumfrence."

Geologists teach us that the earth was once a melted
ball that revolved around the sun blazing forth torrents
of liquid fire. and that after millions of years had
passed away the crust became cold and hardened so as
to sustain animal and vegetable life.   Sir Charles Lyell estimates the internal heat of the earth at a distance
of twenty-five miles from the surface sufficient to melt
the granite rock, and at the distance of thirty-four
miles "to render fluid every known substance."

If the doctrine of the geologists is true, there is
nothing strange in the belief that the "lost island of
Atlantis" once stood above the ocean and was broken
to fragments by the fearful outburst of imprisoned
fire which would tear the land into fragments and ex-
pose it to the maddened waves.

Some of the advocates of this Atlantic theory, especially Brasseur de Bourburg, claim that the civilization of Egypt and other parts of the continent of Europe was carried from the "lost Atlantis," and that the ancient dwellers there were allied to the builders of the pyramids and other great ancient cities which flourished on the Eastern Continent in pre-historic times. In referring to the pyramids this historian says that "for every pyramid in Egypt there can be found a hundred in Central America," hence he would argue that America was peopled long before the foundations of these magnificent structures were laid.

That Egypt, Phœnicia and Syria were peopled by an intelligent and civilized race of men in the far distant past is beyond a question. But whether the civilization originated in America or the "lost Atlantis" will never be known. That Central America's civilization existed prior to that of Egypt is not improbable, though the duration of that of Egypt and Syria points to a vast period of time. On the subject of the "Antiquity of Civilization," M. Oppert, one of the great scientists of Europe, in an essay read before the Brussels Congress, claims to show from the astronomical observations of the Egyptians and Assyrians, "that 11,542 years before our era man existed on the earth at such a state of civilization as to be able to take note of astronomical phenomena, and to calculate with considerable accuracy the length of the year. The Egyptians," says he, "calculated the cycles of 1,460 years— zodiacal cycles, as they were called. Their year consisted of 365 days, which caused them to lose one day

in every four solar years, and consequently they would
attain their original starting point again only after
1,460 years (365 × 4). Therefore, the zodiacal cycle
ending in the year 139 of our era commenced in the
year 1,322 B. C. The Chaldeans state that between
the deluge and their first historic dynasty there was a
period of 39,180 years. Now, what means this number? It stands for 12 Egyptian zodiacal cycles *plus*
12 Assyrian lunar cycles. On the other hand, the Assyrian cycle, 1,805 years, or 22,325 lunations. An Assyrian cycle begun 712 B. C.

$$\left. \begin{array}{l} 12 \times 1.460 = 17,520 \\ 12 \times 1,805 = 21,660 \end{array} \right\} = 39,180.$$

"These two methods of calculating time are in agree-
ment with each other, and were known simultaneous-
ly to one people—the Chaldeans. Let us now build up
the series of both cycles starting from our era, and the
result will be as follows:

ZODIACAL CYCLE	LUNAR CYCLE
1,460	1,805
1,322	712
2,782	2,517
4,242	4,322
5,702	6,127
7,162	7,932
8,622	9,737
10,082	11,542
11,542	

"At the year 11,542 B. C. the two cycles came together and consequently they had on that year their
common origin in one and the same astronomical observation."

THE CIVILZATION PECULIAR TO AMERICA.

Whatever theory may be adopted with regard to the ti.ne when the ancient inhabitants of America whose foot-prints only can be seen, arrived to the stage of civilization which their works would seem to indicate, one thing is sure that their monuments differ from the works of any other people, and "are entirely and abso- lutely anonomalous—they stand alone." The great works found in Central America are, in the opin- ion of many investigators, older that any of the cities of Egypt, Syria, Phœnicia or Southern Asia.

It is not known whether the ancestors of the ancient Mound Builders, Mexicans, Central Americans or those who built the great structures in South America first saw the light on the Western or the Eastern Continent, but one thing cannot be denied and that is that they descended from a race of naked savages elevated but little above the denizens of the forest, and that it re- quired long ages of time to produce the builders of Palenque, Copan, Mitala, Mayapan and the other great str.ictures (now in ruins) found in Central America.

If the work of the Mound Builders and. other an- cient people were constructed some thousands of years ago the length of time required to attain such a re- markable civilization since the infancy of the race must have been far away in the night of the past. The savage state of the primitive man is attested by his residence in dens and caves devouring shell fish, and

contending for existance with the huge monsters of
the rivers, and forests; and then slowly emerging
from this low and degraded condition by the invention
of impliments composed of flint and bone and thence
to higher conditions until he culminated in the wonderful civilization of the builders of magnificent cities.
But the time consumed through long geological periods must have been immense, and then again that
the wonderful ruins now wasting away in South and
Central America are not the first great cities constructed there, for evidence is abundant that they
have been several times rebuilt.

After giving this subject a thorough examination, I
am led to adopt the theory that the civilization which
thousands of years ago constructed such elegant buildings, such grand and imposing pyramids, originated
here in America. Notwithstanding we find emblems
which have some resemblance to those once used by
the ancient Phœnicians, Egyptians and Assyrians, that
fact fails to prove a common origin.

That numerous types of human beings have occupied some portions of this earth of ours for untold ages
is well established, and also that they differ in many
particulars. The nations of Africa who fish in the
Nile and contend with the alligator, the lion and tiger
for supremacy, can always be distinguished from every
other race of men by their woolly hair, retreating for-
head, thick lips and ebony skin. The Indian tribes of
America by nature are savage and possess a complex-
ion and disposition which differs from any other peo-

ple. and though he should pass through a thousand generations he still would retain his characteristics. The same would hold true of the Negro, for he could never attain the dark silken locks of the Indian or his cruel and savage nature. Whether they should live on the sands of Arabia, on the islands in the ocean, or on the inhabitable portion of either continent, they will ever retain their identity. "The leopard can never change his spots nor the Ethiopian his skin." The tall man has an ancestry and brotherhood of tall men. "The Syrian nose of the Jew is on the monuments of Egypt carved three thousand years ago."

I do not contend that the Indian of America may not become modified to some extent by mixing with other types, but I do contend that he could never attain to the civilization, genius and energy possessed by the builders of the pyramids and ancient cities found in Central America. It is now near three hundred years since various Indian tribes have been exposed to the civilization of the European displayed in the mechanic arts and the preaching of the gospel of Christ. But notwithstanding the vast labor and money which has been expended to teach him, he builds no pyramids, no cities, but seems demoralized by every effort made for his civilization.

That man is a progressive being is an established fact, and that the American Indian, the Negro of Africa and the other tribes of uncivilized men now occupying many portions of the earth were once much lower in the intellectual scale than they are at present is very probable. But nature has placed bounds to progress of every kind; "men can never be angels nor angels Gods." The trees of the forest present different types and have obeyed the law of progression for untold ages, yet the beech, the maple and the oak can never reach the height and magnitude of the matchless trees of California who tower near four hundred feet towards the clouds and reach a circumference of more than a hundred. Notwithstanding all of the various types of men commenced low down in the scale there has been but few so constructed by nature as to ever attain to the intellect of a Newton, a Leibnitz, or the sages of Greece and Rome.

With regard to the builders of those ancient cities in Central America, I believe they spring from a race of men different from any other types, and who possessed in their nature the elements of progress far superior to the Indian or Negro; and, that in passing through hundreds of generations they attained the knowledge, energy and ability, to construct important cities.

The general appearance of the ancient works found in Central and South America, and the vast period of time which they indicate, being worn and broken to fragments by the action of the elements, has led many antiquarians to adopt the theory that America is the birthplace of the earliest race of man. Whether this king of created beings first saw the glorious sun or the brilliant orbs that revolve around him on the American Continent or not, one thing is probable, and that is, that the first civilization represented by the ancient people originated here, and very likely in the vicinity where it is found.

The advocates of this American theory claim that the Western Continent is geologically much older than that of Europe, and was peopled with various forms of life ages before the Eastern Continent was raised from beneath the ocean, and that man originated either in South or Central America; or it might have been the soil of the lost Atlantis which was first trodden by human feet.

THE END.

www.ingramcontent.com/pod-product-compliance
Lightning Source LLC
Chambersburg PA
CBHW021032210326
41598CB00016B/990